Technician's Guide

Programmable Controllers

Technician's Guide to
Programmable Controllers
Third Edition

Richard A. Cox
Spokane Community College

Delmar Publishers

I(T)P An International Thomson Publishing Company

Albany • Bonn • Boston • Cincinnati • Detroit • London • Madrid • Melbourne
Mexico City • New York • Pacific Grove • Paris • San Francisco • Singapore • Tokyo
Toronto • Washington

NOTICE TO THE READER

Publisher does not warrant or guarantee any of the products described herein or perform any independent analysis in connection with any of the product information contained herein. Publisher does not assume, and expressly disclaims, any obligation to obtain and include information other than that provided to it by the manufacturer.

The reader is expressly warned to consider and adopt all safety precautions that might be indicated by the activities herein and to avoid all potential hazards. By following the instructions contained herein, the reader willingly assumes all risks in connections with such instructions.

The publisher makes no representation or warranties of any kind, including but not limited to, the warranties of fitness for particular purpose of merchantability, nor are any such representations implied with respect to the material set fourth herein, and the publisher takes no responsibility with respect to such material. The publisher shall not be liable for any special, consequential, or exemplary damages resulting, in whole or part, from the readers' use of, or reliance upon, this material.

Cover photo courtesy of Allen-Bradley Co., Inc., Eaton Corporation, Modicon, Inc.

DELMAR STAFF
Senior Editor: Mark W. Huth
Editorial Assistant: Michelle Ruelos Cannistraci
Project Editor: Eleanor Isenhart
Production Coordinator: Dianne Jensis
Art/ Design Coordinator: Heather Brown

Copyright © 1995
By Delmar Publishers
a division of International Thomson Publishing Inc.

The ITP logo is a trademark under license

Printed in the United States of America

For more information, contact:
Delmar Publishers
3 Columbia Circle, Box 15015
Albany, New York 12212-5015

International Thomson Publishing Europe
Berkshire House 168-173
High Holborn
London WC1V7AA
England

Thomas Nelson Australia
102 Dodds Street
South Melbourne, 3205
Victoria, Australia

Nelson Canada
120 Birchmount Road
Scarborough, Ontario
Canada M1K 5G4

Delmar Publishers' Online Services

To access Delmar on the World Wide Web, point your browser to:
http://www.delmar.com/delmar.html

To access through Gopher: gopher://gopher.delmar.com
(Delmar Online is part of "thomson.com", an Internet site with information on more than 30 publishers of the International Thomson Publishing organization.)

For information on our products and services:
email: info@delmar.com
or call 800-347-7707

International Thomson Editores
Campos Eliseos 385, Piso 7
Col Polanco
11560 Mexico D F Mexico

International Thomson Publishing GmbH
Königswinterer Strasse 418
53227 Bonn
Germany

International Thomson Publishing Asia
221 Henderson Road
#05-10 Henderson Building
Singapore 0315

International Thomson Publishing–Japan
Hirakawacho Kyowa Building, 3F
2-2-1 Hirakawacho
Chiyoda-ku, Tokyo 102
Japan

5 6 7 8 9 10 XXX 01 00 99 98 97 96

Library of Congress Cataloging-in-Publication Data

Cox, Richard A.
    Technician's guide to programmable controllers / Richard A. Cox. —3rd ed.
        p. cm.
    Includes index.
    ISBN 0-8273-6238-2
    1. Programmable controllers. I. Title. II. Title: Programmable controllers.
TJ223.P76C69 1995
629.8'9—dc20

94-24400
CIP

TABLE OF CONTENTS

Preface xi

Chapter 1

What is a Programmable Logic Controller (PLC)? 1
    Chapter Summary 10
    Review Questions 10

Chapter 2

Understanding the Input/Output (I/O) Section 11
    I/O Section 11
    Discrete I/O Modules 16
    Analog I/O Modules 35
    Chapter Summary 43
    Review Questions 44

Chapter 3

Processor Unit 46
    The Processor 47
    Memory Types 52
    Memory Size 54
    Memory Structure 56
    Peripherals 57
    Chapter Summary 59
    Review Questions 60

Chapter 4

Programming Devices (Programmers) 61
    Programming Devices 61
    Personal Computer Programmers 71
    Chapter Summary 72
    Review Questions 72

Chapter 5

Memory Organization 73
    Memory Words and Word Locations 73
    Memory Organization 79
    Allen-Bradley PLC-5 File Structure 84
    Chapter Summary 86
    Review Questions 87

Chapter 6

Numbering Systems 88
    Decimal System 88
    Binary System 90
    Octal System 92
    Hexadecimal System 94
    Binary Coded Decimal (BCD) System 97
    Using Numbering Systems 98
    Chapter Summary 100
    Review Questions 101

Chapter 7

Understanding and Using Ladder Diagrams 102
    Wiring Diagrams 102
    Ladder Diagrams 103
    Ladder Diagram Rules 104
    Basic Stop/Start Circuit 106
    Sequenced Motor Starting 108
    Chapter Summary 109
    Review Questions 110

Chapter 8

Relay Type Instructions 112
    Programming Contacts 113
    EXAMINE ON 116
    EXAMINE OFF 116
    Clarifying EXAMINE ON and EXAMINE OFF 121
    Chapter Summary 124
    Review Questions 125

Chapter 9

Programming a PLC 126
    Sample Programs 127
    Programming with a Computer 138
    Instruction Sets 145
    PLC-5 Addressing Scheme 145
    Peripherals 159
    Chapter Summary 160
    Review Questions 160

Chapter 10

Programming Considerations 161

Network Limitations 161
Programming Restrictions 165
Program Scanning 168
Programming Stop Buttons 170
Logical Holding Instructions 171
Discrete Holding Contacts 171
Overload Contacts 171
Chapter Summary 173
Review Questions 173

Chapter 11

Program Control Instructions 175
Master Control Relay Instructions 175
Latching Relay Instructions 177
Safety Circuit 180
Immediate Input Instruction 181
Immediate Output Instruction 182
Jump and Label Instructions 184
Jump to Subroutine, Subroutine, and Return Instructions ... 185
Temporary End Instruction 185
Always False Instruction 186
One Shot Instruction 186
Chapter Summary 187
Review Questions 187

Chapter 12

Programming Timers 189
Pneumatic Timers (General) 189
Allen-Bradley PLC-2 Timers 195
Allen Bradley PLC-5 Timers 203
Square D Company Timers 211
Modicon Inc. Timers 217
Cascading Timers 220
Chapter Summary 222
Review Questions 222

Chapter 13

Programming Counters 224
Allen-Bradley PLC-5 Counters 227
Modicon 984 Counters 233
Square D Company Counters 236
Combining Timers and Counters 237
Chapter Summary 238
Review Questions 238

Chapter 14

Data Manipulation 241
   Data Transfer 241
   Allen-Bradley PLC-5 Data Transfer Instructions 247
   Data Compare 250
   Allen-Bradley PLC-5 Data Compare Instructions 256
   Chapter Summary 262
   Review Questions 262

Chapter 15

Math Functions 264
   Using Math Functions 264
   Arithmetic Functions Using the Allen-Bradley PLC-2 267
   Allen-Bradley PLC-5 Math Functions 268
   Gould 984 Arithmetic (Math) Functions 270
   2s Complement 274
   Chapter Summary 282
   Review Questions 283

Chapter 16

Word and File Moves 284
   Words 284
   Synchronous Shift Register 284
   File Moves 289
   Word-to-File Instruction 289
   File-to-Word Instruction 290
   File-to-File Instruction 291
   Asynchronous Shift Register (FIFO) 297
   Last In–First Out (LIFO) 301
   Chapter Summary 301
   Review Questions 301

Chapter 17

Sequencers 303
   Masks 306
   Chapter Summary 308
   Review Questions 308

Chapter 18

Programming With Boolean 310
   Boolean Algebra 310
   Programming in Boolean 314
   Timers 323

Defining Small PLCs 324
Chapter Summary 325
Review Questions 325

Chapter 19

Understanding Basic MS-DOS® Commands 328
Starting the Computer 328
Formatting a Disk 329
Creating Files and Directories 330
Directory (DIR) Command 333
Interrupt Command (Break Command) 335
Clear Screen Command 335
Copy Command 335
Making Directories Command 336
Change Directories Command 337
Renaming Files Command 337
Delete Files Command 337
Remove Directory Command 338
Directory Tree Command 338
Chapter Summary 340
Review Questions 340

Chapter 20

Start Up and Troubleshooting 341
Start Up 341
Testing Inputs 343
Testing Outputs 344
Final System Checkout 346
Troubleshooting 347
Chapter Summary 351
Review Questions 351
Glossary 353
Index 365

Preface

The programmable logic controller, first introduced in 1969, has become an unqualified success. Programmable logic controllers, or PLCs as they are often referred to, are now produced by over 50 manufacturers. Varying in size and sophistication, these electronic devices have become the industry standard, replacing the hard-wired electromechanical devices and circuits that have controlled the process machines and driven equipment of industry in the past.

With every major motor control equipment manufacturer, and some computer companies now offering PLCs, it is impossible to write a book that explains how they all work and/or are programmed. Instead, this book is intended to discuss PLCs in a general, or generic sense, and to cover the basic concepts of operation that are common to all programmable logic controllers.

Many electricians and/or technicians seem apprehensive about PLCs and their application in industry. One of the purposes of this text is to explain PLC basics in a plain, easy-to-understand, approach so that electricians and technicians with no PLC experience will be more comfortable with in their first exposure to programmable logic controllers.

Half the battle of understanding any programmable logic controller is to first understand the terminology of the PLC field. This text covers terminology, as well as explaining the input/output section, processor unit, programming devices, memory organization, and much more.

A chapter has been included to explain not only ladder diagrams, but also RELAY LADDER LOGIC, which is the programming language used in the majority of programmable controllers today.

Examples of basic programming techniques with typical PLCs are discussed and illustrated, as well as the commonly used commands and functions. Although this text only scratches the surface of available commands and other advanced programming capabilities of the current PLCs on the market today, the reader will gain a basic understanding of the most commonly used commands and how they work. As with any new skill, a firm base of understanding is required before an electrician or technician can become proficient at that skill. After completing the text, the reader will possess a good foundation upon which additional PLC skills and understanding can be built.

The best teacher, of course, is experience, and as stated several times in the text, the only way to really understand any given PLC is to work with that PLC. If a PLC is not available, the next best thing is a workshop or seminar sponsored by a local PLC distributor. If a workshop or seminar is not available, obtain as much literature and other information as possible from a local electrical distributor or PLC representative.

The PLC manufacturers are reducing prices as well as adding new features and program capabilities every day. With the rapid advancements in PLCs, the electrician without an electronics background need not feel intimidated. The manufacturers are doing everything possible to make the PLC easy to install, program, troubleshoot, and maintain.

The author is a member of the Electrical/Robotics Department at Spokane Community College in Spokane, Washington. He holds a BS degree from the University of the State of New York and a MS degree from Eastern Washington University, and is also a member of the International Brotherhood of Electrical Workers, Local 73.

The author wishes to acknowledge the cooperation and assistance of the different manufacturers whose product information and photographs are used throughout the text to illustrate different concepts and PLC components. All these manufacturers are leaders in the PLC field, and each offers a full line of programmable logic controllers. It is not a question of which PLC is best, but rather which PLC best fits your needs and individual application.

The author would also like to acknowledge the technical assistance provided by Rodney Hedman, an instructor in the Electrical/Robotics Department at Spokane Community College, and reviewers John D. Sprague-Williams, College of Dupage, Illinois, Roy Slagle, Greenville Tech, South Carolina, Rick House, Coosa Valley Tech, Georgia, Larry Kramer, Washtenaw Community College, Michigan, Gary Dunning, Hennepin Technical College, Minnesota, and James Ahneman, Chippewa Valley Technical College, Wisconsin.

Chapter **1**

What is a Programmable Logic Controller (PLC)?

Objectives

After completing this chapter, you should have the knowledge to

- Describe several advantages of a programmable logic controller (PLC) over hard-wired relay systems.
- Identify the four major components of a typical programmable logic controller and describe the function of each.
- Define the term discrete.
- Define the term analog.
- Identify different types of programming devices.

A programmable logic controller is a solid-state system designed to perform the logic functions previously accomplished by components such as electromechanical relays, drum switches, mechanical timers/counters, etc., for the control and operation of manufacturing process equipment and machinery.

Even though the electromechanical relay (control relays, pneumatic timer relays, etc.) has served well for many generations, often, under adverse conditions, the ever-increasing sophistication and complexity of modern processing equipment requires faster acting, more reliable control functions than electromechanical relays and/or timing devices can offer. Relays have to be hard-wired to perform a specific function, and when the system requirements change, the relay wiring has to be changed or modified. In extreme cases, such as in the auto industry, complete control panels had to be replaced since it was not economically feasible to rewire the old panels with each model changeover.

It was, in fact, the requirements of the auto industry and other highly specialized, high-speed manufacturing processes that created a demand for smaller, faster acting, and more reliable control devices. The electrical/electronics industry responded with modular-designed, solid-state electronic devices. These early devices, while offering solid-state reliability, lower power consumption, expandability, and elimination of much of the hard-wiring, also brought with them a new language. The language consisted of AND gates, OR gates, NOT gates, OFF RETURN MEMORY, J-K flip flops, and so on.

What happened to simple relay logic and ladder diagrams? That is the question the plant engineers and maintenance electricians/technicians asked the solid-state device manufacturers. The reluctance of the end user to learn a new language and the advent of the microprocessor gave the industry what is now known as the **programmable logic controller** (PLC). The first programmable logic controller was invented in 1969 by Richard (Dick) E. Morley, who was the founder of the Modicon Corporation.

Internally there are still AND gates, OR gates, and so forth in the processor, but the design engineers have preprogrammed the PLC so that programs can be entered using RELAY LADDER LOGIC. While RELAY LADDER LOGIC may not have the mystique of other computer languages such as FORTRAN and COBOL, it is a high-level, real-world, graphic language that is understood by most electricians.

The National Electrical Manufacturing Association (NEMA) defines a programmable controller as follows:

> A programmable controller is a digital electronic apparatus with a programmable memory for storing instructions to implement specific functions, such as logic, sequencing, timing, counting, and arithmetic to control machines and processes.

What does a PLC consist of, and how is it different from a computer control system? The PLC consists of a programming device (keyboard), processor unit, power supply, and an input/output **interface** such as the computer system illustrated in Figure 1–1. And while there are similarities, there are also some major differences.

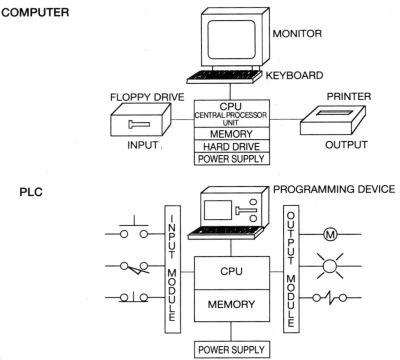

Figure 1–1 Comparison of a Computer System and a PLC

NOTE: An interface occurs when two systems come together and interact, or communicate. In the case of the PLC, the communication or interaction is between the inputs (limit switches, push buttons, sensors, etc.), outputs (coils, solenoids, lights, and so forth), and the processor. This interface happens when any input voltage (AC or DC) or current signal is changed to a low-voltage DC signal that the processor uses internally for the decision-making process.

PLCs are designed to be operated by plant engineers and maintenance personnel with limited knowledge of computers. Like the computer, which has an internal memory for its operation and storage of a program, the PLC also has memory for storing the user program, or LOGIC, as well as a memory for controlling the operation of a process machine or driven equipment. But unlike the computer, the PLC is programmed in RELAY LADDER LOGIC, *not* one of the computer languages. It should be stated, however, that some PLCs will use a form of Boolean Algebra to enter the RELAY LADDER LOGIC. A brief description of Boolean Algebra will be covered in Chapter 18.

The PLC is also designed to operate in the industrial environment with wide ranges of ambient temperature, vibration, and humidity, and is not usually affected by the electrical noise that is inherent in most industrial locations.

NOTE: Electrical noise is discussed in Chapter 2.

Maybe one of the biggest, or at least most significant, differences between the PLC and a computer is that PLCs have been designed for installation and maintenance by plant electricians who are not required to be highly-skilled electronics technicians. Troubleshooting is simplified in most PLCs because they include fault indicators, blown-fuse indicators, input and output status indicators, and written fault information that can be displayed on the programmer.

Although the PLC and the computer are different in many ways, the computer is often used for programming and monitoring the PLC. Using computers in conjunction with PLCs will be discussed in later chapters.

A typical PLC can be divided into four components. These components consist of the **processor unit**, **power supply**, the **input/output section** (interface), and the **programming device**.

The **processor unit** houses the processor which is the decision-maker, or "brain" of the system. The brain is a microprocessor-based system that replaces control relays, counters, timers, sequencers, and so forth, and is designed so that the user can enter the desired program in RELAY LADDER LOGIC. The processor then makes all the decisions necessary to carry out the user program, based on the status of the inputs and outputs for control of a machine or process. It can also perform arithmetic functions, data manipulation, and communications between the local input/output section, remotely located I/O sections, and/or other networked PLC systems. Figure 1–2 shows an Allen-Bradley PLC-5/40 processor unit.

NOTE: Some manufacturers refer to the processor as a **CPU** or central processing unit.

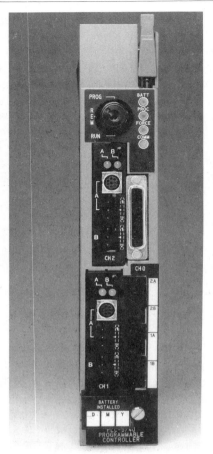

Figure 1–2 Allen-Bradley
PLC-5/40 Processor Unit
(Courtesy of Allen-Bradley)

The **power supply** is necessary to convert 120 or 240 volts (V) AC voltages to the low voltage DC required for the logic circuits of the processor, and for the internal power required for the I/O modules. The power supply can be a separate unit as shown in Figure 1–3, one of modular design that plugs into the processor rack as shown in Figure 1–4, or one, depending on the manufacturer, that is an integral part of the processor.

NOTE: The power supply does not supply power for the actual input or output devices themselves; it only provides the power needed for the internal circuitry of the input and output modules. DC power for the input and output devices, if required, must be provided from a separate source.

Figure 1–3 Allen-Bradley Separate Power Supply *(Courtesy of Allen-Bradley)*

Figure 1–4 Modular (Plug-in) Power Supply *(Courtesy of GE Fanuc Automation)*

The power supply can be broken down into four basic parts as shown in Figure 1–5. The first block, or section, of the power supply consists of a step-down transformer. The step-down transformer reduces the voltage level of the incoming AC power. Many power supplies use a step-down transformer that is also a constant voltage transformer. A constant voltage transformer maintains a constant output voltage, even if the incoming power is fluctuating. The second por-

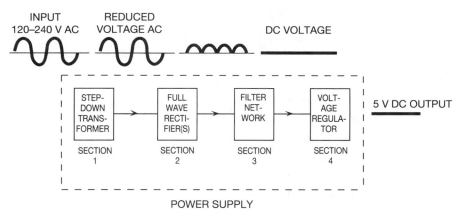

Figure 1–5 Block Diagram of a Typical Power Supply

tion of the power supply is the rectifier section, and contains the full wave bridge rectifier(s) to convert the AC sine wave from the secondary of the transformer to a pulsating DC voltage (shown by the wave form in Figure 1–5). The pulsating DC voltage must be further conditioned before it can be used by the processor and I/O modules. The third section of the power supply, the filter section, uses filter devices and/or networks to filter and smooth the DC voltage coming from the rectifier section. The final section of the power supply consists of a voltage regulator. The regulator's function is to maintain a constant DC output voltage, even if the incoming AC voltage fluctuates or varies due to load changes or line disturbances.

The size or amperage rating of the power supply is based on the size, number, and type of I/O modules that are to be used. Power supplies are normally available with output current ratings of 3–20 amps.

NOTE: Consider future needs and the possibility of expansion when initially sizing the power supply. It is cheaper in the longrun to install a larger power supply initially than to try and add additional capacity at a later date.

The **input/output section** consists of input modules and output modules. The number of input and output modules necessary is dictated by the requirements of the equipment that is to be controlled by a PLC. Figure 1–6 shows an input/output section. Modules are "plugged-in" or added as required.

Input and output modules, referred to as the I/O (I for input and O for output) are where the real-world devices are connected. The real-world input (I) devices can be push buttons, limit

Figure 1–6 Inserting a 32-Point Input Module into a
Modicon I/O Rack
(Courtesy of Modicon Inc.)

switches, analog sensors, thumbwheels, selector switches, etc., while the real-world output devices (O) can be hard-wired to motor starter coils, solenoid valves, indicator lights, positioning valves, and the like. The term *real world* is used to separate actual devices that exist and must be physically wired as compared to the internal functions of the PLC system that duplicate the function of relays, timers, counters, and so on, even though none physically exists. This may seem a bit strange and hard to understand at this point, but the distinction between what the processor can do internally—which eliminates the need for all the previously-used control relays, timers, counters, and so forth—will be graphically shown and readily understandable later in the text.

Real-world input and output devices are of two types: discrete and analog. *Discrete* I/O devices are either *ON* or *OFF*, *open* or *closed*, while *analog* devices have an infinite number of possible values. Examples of analog input devices are temperature probes, and pressure indicators. The input from an analog input device (varying voltage or current) is converted by way of an Analog-to-Digital Converter (ADC). The conversion value is proportional to the analog input signal and is stored in memory for later use or comparison. Limit switches, push buttons, and the like, are examples of *discrete input devices*.

Examples of *discrete output devices* are motor starter coils, solenoids, and indicator lamps. Analog output devices require varying voltage or current levels to control the analog output. The varying value will be accomplished using a Digital-to-Analog Converter (DAC). The analog output normally is isolated from the output logic circuitry by means of optical coupling.

NOTE: Optical coupling will be discussed in detail in Chapter 2.

A reference was made earlier in this chapter to the I/O section as an interface. Although not a common reference, it is an accurate one. The I/O section contains the circuitry necessary to convert input voltages of 120–240 V AC or 0–24 V DC, etc., from discrete input devices to low-level DC voltages (typically 5 V) that the processor uses internally to represent the status or condition (*ON* or *OFF*). The I/O section can also convert 4–20 milliampere (mA) input signals to low-level DC voltages for the processor. Similarly, the output module changes low-level DC signals from the processor to 120–240 V AC or DC voltages required to operate the discrete output devices. This is a brief overview of the I/O section and its function. How input and output devices are wired to I/O modules and more information about the module circuitry itself is covered in Chapter 2.

The **programming device** is used to enter the desired program or sequence of operation into the PLC memory. The program is entered using RELAY LADDER LOGIC, and it is this program that determines the sequence of operation and ultimate control of the process equipment or driven machinery. The programming device can be any one of three types: hand-held; dedicated; or personal computer.

The hand-held programmer uses either **LED** (light emitting diode) or **LCD** (liquid crystal display). The hand-held programmers are small, lightweight, and convenient to use in the field (Figure 1–7). The small size, however, limits the display capabilities, which in turn limits its effectiveness for reviewing the program or using the programmer for troubleshooting.

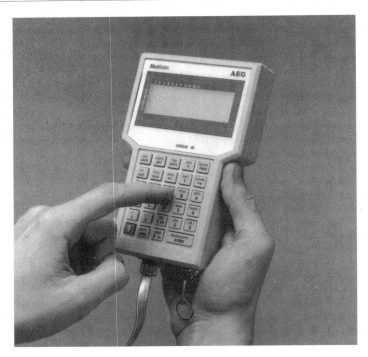

Figure 1–7 Hand-held Programmer
(Courtesy of Modicon Inc.)

Several manufacturers offer a dedicated programmer that uses a standard video display terminal (**VDT**) for displaying the program. The use of a standard viewing screen allows more of the program to be viewed at one time and makes troubleshooting and memory access much easier. A dedicated programming terminal is shown in Figure 1–8.

A personal computer (PC) can be used to program most of the PLCs on the market today if it can run DOS® and is IBM® compatible. Some PLCs require only software to communicate with a personal computer, while others require special hardware keys and/or communication cards for them to work successfully as programming devices. Once communications between the personal computer and the PLC have been established, the PC provides all the benefits of the dedicated programmer, plus it provides program storage as well as runs all the various software packages we have come to depend on today such as spreadsheets, word processing, and graphics. Due to the versatility and lower cost—when compared to dedicated programmers—PCs have become the most popular programming device. Figure 1–9 shows a laptop personal computer that, with the appropriate software, is used to control a programmable controller.

Figure 1–8 Dedicated Progammer
(Courtesy of Allen-Bradley)

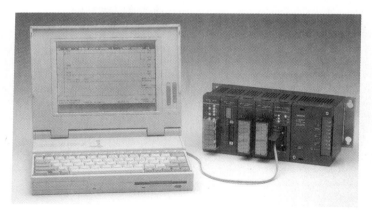

Figure 1–9 Laptop Computer Connected to a Siemans
SIMATIC TI305
(Courtesy of Siemens Industrial Automation)

Chapter Summary

Programmable logic controllers (PLCs) have made it possible to precisely control large process machines and driven equipment with less physical wiring and lower installation costs than is required with standard electromechanical relays, pneumatic timers, drum switches, and so on. The programmability allows for fast and easy changes in the RELAY LADDER LOGIC to meet the changing needs of the process or driven equipment without the need for expensive and time-consuming rewiring. By designing the modern programmable logic controller (PLC) to be "technician friendly," the PLC is easier to program and be used by plant engineers and maintenance technicians who have little or no electronic background.

REVIEW QUESTIONS

1. List the four main components of a programmable logic controller.
2. Define the term *interface*.
3. Define the term *real world*.
4. Define the term *discrete*.
5. Define the following initials or acronyms:

LED	PLC	NEMA
CPU	ADC	LCD
VDT	DAC	PC

6. Define the term *analog*.
7. List the three types or styles of programming devices.
8. T   F   RELAY LADDER LOGIC is a high-level graphic computer language.
9. What is the major advantage of a PLC system over the traditional hard-wired control system?
10. Draw a block diagram of a typical DC power supply.

Chapter 2

Understanding the Input/Output (I/O) Section

Objectives

After completing this chapter, you should have the knowledge to
- Describe the I/O section of a programmable controller.
- Identify DIP switches.
- Describe how basic AC and DC input and output modules work.
- Define *optical isolation* and describe why it is used.
- Describe the proper wiring connections for input and output devices and their corresponding modules.
- Explain why a hard-wired emergency-stop function is desirable.
- Define the term *interposing*.
- Describe what I/O shielding does.
- List environmental concerns when installing PLCs.

I/O Section

The input/output section, or I/O section, is the major reason that PLCs are so versatile when used with process machines or driven equipment. The I/O section has the ability to change virtually any type of voltage or current signal into a logic-level signal (typically 5 V DC) that is compatible with the processor. The I/O section automatically makes the conversions necessary for the processor to interpret input signals and to activate output devices, even when the input and output devices are of various voltage and current levels.

A DC input module, for example, can be used with a 12 V DC proximity switch to turn on a 240 V AC motor starter coil that is connected to an AC output module. The conversion and interfacing is all accomplished automatically in the I/O section of the PLC, and it is this ease with which the interfacing is accomplished that has made the PLC such a viable tool in industrial and process control.

The input modules of the I/O section provide the status (*ON* or *OFF*) of push buttons, limit switches, proximity switches, and the like, to the processor so decisions can be made to control the machine or process in the proper sequence. Outputs, such as motor-starter coils, indicator lights, and solenoids are interfaced to the processor through the output section of the I/O. Once a decision has been made by the processor, a signal is sent to the output section to control the flow of current to the output device. In general, the status of the inputs are relayed to the processor and based on the logic of the program that has been written, a decision is made to turn the outputs to *ON* or *OFF*. All of the different types and levels of signals (voltages and currents) used in the control process are interfaced in the I/O section.

The I/O section generally can be divided into two categories: **fixed I/O** and **modular I/O**.

Fixed I/O

PLCs with fixed I/O typically come in a complete unit that contains the processor, I/O section, and power supply. The I/O section contains a fixed number of inputs and outputs, all of which have the same voltage level (120 V AC, 24 V DC or 230 V AC). For example, the Modicon Micro shown in Figure 2–1 has 16 inputs and 12 outputs in the self-contained base unit. The base unit measures 10 inches long by 5 inches high and is only 3 inches deep. Like most small PLCs, the Modicon Micro can be DIN-rail or panel mounted. Figure 2–1a shows the DIN–rail mounting instructions for the GE Fanuc Micro PLC.

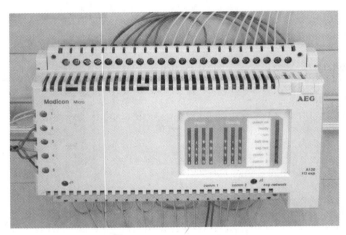

Figure 2–1 Modicon Micro PLC
(Courtesy of Modicon Inc.)

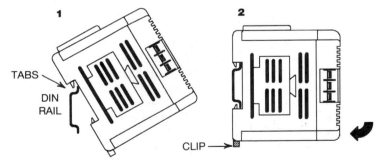

Position the upper edge of the unit over the DIN rail, so that the rail is behind the tabs as shown above.

Pivot the unit downward (for a unit being mounted right side up) until the spring-loaded clip in the bottom of the unit clicks firmly into place.

Figure 2–1a DIN Rail Mounting
(Courtesy of GE Ganuc Automation)

If more I/O capability is required or different voltages are needed, expansion units with various I/O configurations can be added (see Figure 8–1a).

Small PLCs with fixed I/O typically have a discrete input and output section. As discussed in Chapter 1, discrete-type I/O signals are *ON* or *OFF* and do not vary in level. When a 120 V limit switch closes or is *ON*, the signal to the input section will be 120 V and 0 V when the switch is open (*OFF*). Discrete I/O is sometimes referred to as digital I/O, whereas digital signals are either *ON* or *OFF*, and do not vary in magnitude once the signal is established. An analog-type signal, on the other hand, will vary in magnitude and be constantly changing based on control variables in the process. A pressure transducer that sends a 20-milliampere signal at 500 psi or a 10-milliampere signal at 250 psi is an example. Many manufacturers offer analog expansion units that can be added to their fixed I/O PLCs.

While these PLCs are small in size, they are big on features. Many include full-feature instruction sets that include timers, counters, sequencers, shift registers, word moves, data compare, and much more. One should consult the specific dealer for a full list of features.

As the cost of these compact units have decreased, their use has increased. The costs are so competitive that any control processes that use only a small number of relays and/or timers can now be accomplished using a small PLC. The use of a small PLC not only saves money, but also gives added reliability and flexibility.

Because of their shape and size, the term "shoebox" or "brick" is often used by manufacturers and users alike when referring to PLCs with fixed I/O. Figure 2–2 shows the Modicon Micro (brick) and the expandable Modicon A120. Note the relative size compared to the ball-point pen.

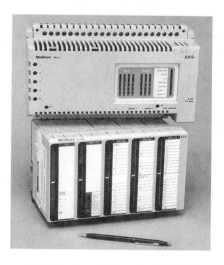

Figure 2–2 Modicon Micro and A 120 PLCs
(Courtesy of Modicon Inc.)

Modular I/O

Modular I/O, as the name implies, is modular in nature, more flexible than fixed I/O units, and provides added versatility when it comes to the type and number of input and output devices that can be connected to the system. The various types of input and output modules that make up the I/O section are housed, or installed, in an I/O rack or chassis.

The I/O rack or chassis is a framework or housing into which modules are inserted. Figure 2–3 shows three different size racks. Figure 2–4 shows the rack with the I/O modules installed and the processor ready for installation.

Figure 2–3 I/O Racks
(Courtesy of Modicon Inc.)

Figure 2–4 I/O Rack with Processor and I/O Modules Installed
(Courtesy of Modicon Inc.)

Racks or chassis come in many shapes and sizes, and typically allow 4, 8, 12 or 16 modules to be inserted. Racks that contain I/O modules and the processor are referred to as *local* I/O. Racks that contain I/O modules, remote I/O communication cards, power supplies, and are mounted separately or away from the processor are referred to as *remote* I/O. An advantage of remote I/O racks is that they can be mounted up to 10,000 feet away from the processor. The number of remote I/O racks that a processor can control varies with each manufacturer. The communication between the remote rack and the processor is accomplished using several different types of communication methods. These methods include coaxial cable, twin axial cable, shielded-twisted pair, or fiber optics. If distance or electrical noise are considerations, the fiber optic communication method may be the best option. Figure 2–5 shows a local rack and three remote I/O racks.

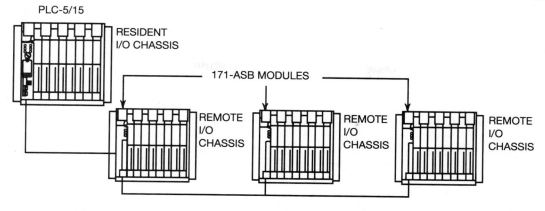

Figure 2–5 Local Rack with Processor and Three Remote I/O Racks
(Courtesy of Allen Bradley)

Whether local or remote, racks normally have jumpers or switches that have to be set or configured in order for the racks to communicate with the processor. A common switch used for rack configuration is referred to as a **DIP** switch. DIP is short for **D**ual-**In**-Line **P**ackage, a common electronic package design, or style, for use on printed circuit boards. These DIP switches are either *ON* or *OFF*, and when set in the proper sequence, are used to assign an address to a rack, such as Rack 1, Rack 2, Rack 3, etc. DIP switches are also used to set fault parameters as well as other processor functions. DIP-switch settings will be specified by the PLC manufacturers and are found in the installation manual.

NOTE: Under no circumstance should a pencil be used to change a DIP-switch position. The graphite in the pencil tip can break off, causing the switch to short. Instead, use the tip of a ballpoint pen or other nonconducting pointed object to change switch positions.

DIP switches are generally mounted on a printed circuit board located in the back of the I/O rack or chassis. This printed circuit board is often referred to as the *backplane*. Figure 2–6 shows a backplane printed circuit board and a DIP-switch group assembly.

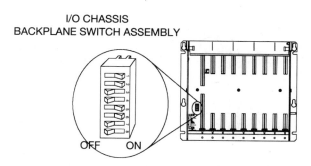

Figure 2–6 Backplane and I/O DIP Switches
(Courtesy of Allen-Bradley)

Within each rack, individual input and output device connections must have a distinct **address** so the processor knows where the device is located, and in return, can send and receive signals, enabling the processor to monitor and/or control the device. Allen-Bradley, for example, uses the rack number, location of a module within a rack, and the terminal number of a module to which an input or output device is connected to determine the device's address. Addresses and addressing of input and output devices will be covered in Chapter 6.

Also mounted on the backplane of the I/O rack are prewired slots or connectors into which the individual I/O modules are inserted. When inserting the modules, proper alignment is assured by card guides (also referred to as printed circuit board guides) which are mounted on the top and bottom of the I/O rack as seen in Figures 2–3 and 2–4.

Input and output modules can be separated into three basic groups: discrete or digital input/output modules; analog input/output modules; and specialized modules.

Discrete I/O Modules

Discrete I/O modules are types of modules that only accept digital or *ON*- and *OFF*-type signals. These modules only recognize these two states or conditions and that again is *ON* or *OFF*. If a discrete device, such as a limit switch, is connected to this type of module, the module determines the state, or position, of the limit switch, and communicates the state, or status, to the processor. If the limit switch is open (*OFF*), the module indicates to the processor that the limit switch is *OFF*. This *OFF* condition is stored in the processor memory as a zero (0). Had the limit switch been in a closed position, the module would have sent a signal to the processor indicating that the limit switch was *ON*, or closed. The *ON* condition would have been stored in the processor memory as a one (1). All information stored in the processor memory about the status or condition of discrete I/O devices are always in ones and zeros.

Discrete modules are the most common type used in a majority of PLC applications and can be divided into two groups: input and output.

Discrete Input Module

The discrete input module communicates the status of the various real-world input devices connected to the module (*ON* or *OFF*) to the processor.

NOTE: The term *real world* is used to indicate that an actual device is involved. As you will learn later in the text, the PLC has the ability to provide timing and counting functions for a machine, even though the timers and counters exist only within the processor, and are not wired into the circuit as with real-world, or actual devices.

Once the real-world input device is connected, an open or closed electrical circuit exists, depending on the position (open or closed) of the device. The status of the real-world input device is then converted to a logic-level DC electrical signal by the input module and sent to the processor.

Discrete input modules come in a wide range of voltages for various applications. Some of the more common voltage modules are 120 V AC, 240 V AC, 24 V DC, and 12–24 V DC. Some manufacturers give their modules on AC/DC rating to increase their flexibility and reduce required inventory. It is important to note, however, that while the module may be used with either AC or DC input voltages, the voltages *cannot* be intermixed on the same module.

Input modules can be purchased with a wide range of input terminals or points, that determine the number of individual field devices that can be connected to the module. Common sizes, depending on the manufacturer, are 4, 8, 16 and 32 points. Sixteen- and 32-point modules are often referred to as high-density modules since they are physically the same size as an 8-point module. High-density modules usually provide lower cost per point, or input device, but are also more difficult to wire. The increased difficulty in wiring is caused by the closer proximity of the wiring terminals and the increased number of wires in the wiring harness.

AC Discrete Input Modules

Figure 2–7 shows a simplified diagram of one of the input circuits of a typical AC discrete input module. Resistors are used to drop the incoming voltage; then a bridge rectifier is used to convert the AC input voltage to DC. Next, a filtering circuit is used to condition the DC and guard against electrical noise. Electrical noise can cause a short-duration DC pulse that is sometimes interpreted by the processor as a closed signal. This false, or erroneous, signal could be interpreted as a valid signal, and a 1 would be placed in memory to indicate the device was *ON*, even though it was not. To eliminate the possibility of faulty operation due to electrical noise, the filter section of the module delays an actual input signal from being sent to the processor for 15 to 25 milliseconds (msec). The filter requires that the AC signal be not only of a specific value, but also be present for a specific amount of time before the module views it as a real signal and communicates the results to the processor. A valid signal is relayed through an **optically-coupled** circuit, across the backplane of the I/O rack to the processor.

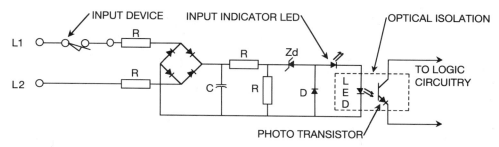

Figure 2–7 Simplified AC Input Module Circuit with Indicator Light

The optically-coupled circuit uses a LED (light emitting diode) to turn *ON*, or forward bias, a photo transistor to complete the electrical circuit to the processor. When the LED is turned *ON* to indicate that the actual input device has closed, the light from the LED is picked up by the photo transistor that makes the transistor conduct, completing a 5 V DC logic circuit, and the status of the input is communicated to the processor. This form of optical coupling is also referred to as **optical isolation**. By employing optical coupling, or isolation, there is no actual electrical connection between the input device and the processor. This eliminates any possibility of the input line voltage, i.e. 120 or 240 V AC, from coming in contact with and damaging the low-voltage DC section of the processor. Optical isolation also protects the processor from electrical noise, voltage transients, or spikes. In summary, optical isolation prevents any unwanted voltage from the I/O section from reaching the logic section of the processor.

Individual status lights are provided for each device that is connected to an input terminal (Figure 2–7). The status light is lit when the input device is closed and is *OFF* when the input device is *OFF* (open). With the status lights showing the actual position of the various input devices connected to the input module, they make a valuable troubleshooting aid. The electrician or technician need only look at the status lights on the input module to determine the position, or status, of any input device.

A typical I/O module consists of two parts: a printed circuit board and a terminal assembly. The printed circuit board plugs into a slot, or connector, in the I/O rack and contains the solid-state electronic circuits that interface the I/O devices with the processor. The terminal assembly then attaches to the front edge of the printed circuit board, which may or may not have a protective cover, depending on the manufacturer. Figures 2–8a and 2–8b show typical input modules.

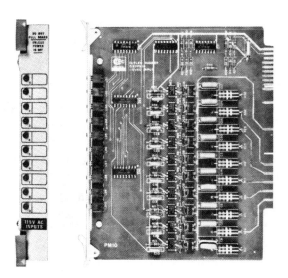

Figure 2–8a 115V AC Input Module
(Courtesy of Cutler-Hammer)

1. Red identification label
2. Status indicators
3. Protective covers
4. Field Wiring Arm connects here
5. Labels identify user inputs
6. Slotted for I/O slot insertion only

Figure 2–8b AC Input Module
(Courtesy of Allen-Bradley)

Figure 2–9 shows how the input module (Figure 2–8b) is installed in the I/O rack.

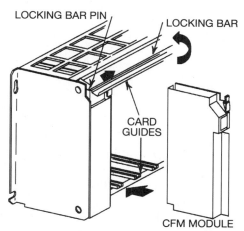

Swing the chassis locking bar down into place to secure the modules. Make sure the locking pins engage.

Figure 2–9 Installing a Module in an I/O Rack
(Courtesy of Allen-Bradley)

After the input modules have been installed in the I/O rack, they are ready to have one side of each input device connected to their terminals or wiring arms (Figure 2–10).

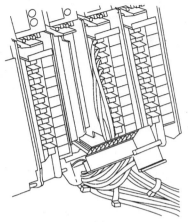

Figure 2–10 I/O Module Field Wiring Arm
(Courtesy of Allen-Bradley)

While each input device has two wires connected, only one wire is connected directly to the input module. The other wire of each input device is connected common to Line 1 (Figure 2–11a). The same connection scheme is used for 16- or 32-point input modules. Figure 2–11b shows the wiring connections for an Allen-Bradley 16-point input module.

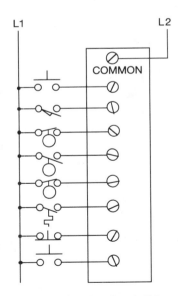

Figure 2–11a Field Wiring for AC Input Module

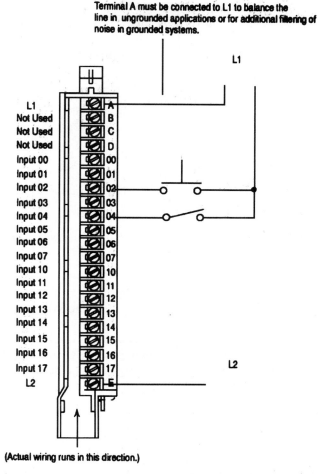

Figure 2–11b Connections for an Allen-Bradley 16-Point Input Module
(Courtesy of Allen-Bradley)

The wires from the individual devices are referred to as field wiring, since the wires are external to the PLC and are connected in the field rather than at the factory. On larger process machines, the field wiring that is brought into the I/O rack consists of hundreds of wires. Because of the number of wires, the wiring is simplified by using a PLC. The basic rule is that one side of each input device is wired to a hot conductor (L1 for AC or + for DC), and the other side of the device is wired to an input terminal on the input module. The input module has a common connection for the neutral, or grounded potential (L2), for AC modules and the negative (–) for DC modules. Consult the literature that comes with each input module to ensure that the correct wiring connections are made.

Figure 2–12 shows two simplified circuits. In the first, or traditional circuit, the input device (single-pole switch) is connected to, and controls, the light. In the PLC circuit, the input device is connected to the input module instead of the light. The module converts the 120 V AC input signal to 5 V DC, and communicates the status of the single-pole switch to the processor.

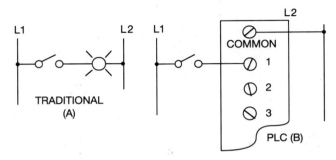

Figure 2–12 Traditional Wiring for Single Pole Switch (A)
Compared to PLC Wiring (B)

DC Discrete Input Module

Figure 2–13 shows a simplified diagram for a typical DC input module. With the exception of the bridge rectifier used in the AC module, the principles are the same. Resistors are used to lower, or drop, the incoming voltage. Filtering circuits condition the low-voltage signal and add an additional delay in the response time. This period of delay is slightly less than the delay used in the AC input module, but it is also used to verify that the signal received is a valid signal of a proper duration, and not a signal caused by electrical noise, voltage transients, or the like.

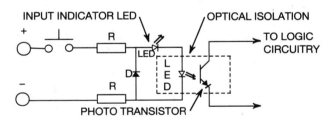

Figure 2–13 Simplified Circuit for DC Input Module with Indicator Light

Optical isolation is also used on the DC input module to isolate the processor from the higher voltage of the input devices. When the LED is turned *ON* by the closing of the input device, the photo transistor conducts and the status of the input device is communicated across the backplane of the I/O rack to the processor. Status lights, shown in the diagram, are also provided for each input device to indicate whether the input device is open or closed. Figure 2–14 shows how a typical DC input module is wired.

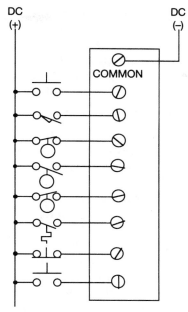

Figure 2–14 Field Wiring for DC Output Module

INPUT

Fast-Responding DC Input Modules

Fast-responding DC input modules are used when the process requires fast-acting sensors to respond to high-speed and/or high-volume applications. Encoders, other types of sensors that respond with many pulses for each rotation of a shaft, or proximity switches that are counting parts or products produced by high-speed machines are all examples of the benefits of a fast-responding module. The internal operation of the module is the same as the discrete DC input module, with the exception of the delay created in the filtering circuit. The fast-responding module has only a 1-msec or less delay. With this short time, normal mechanical contact devices may not work correctly due to the contact bounce that is inherent in some mechanical switches and contacts. The contact bounce is counted as an additional input signal and processed by the processor. This extra count has an obvious adverse affect on machine operation and is not a proper application of this type of input module.

Discrete input modules come in a variety of types and styles that fit most applications requiring an *ON-* or *OFF*-type of signal. Allen-Bradley, for example, manufactures nearly 20 different types of discrete input modules. Selecting the proper module for any given application is relatively easy when the voltage level, voltage type, current, and response time are known.

Discrete Output Modules

The purpose of a discrete output module is to control the current flow to real-world devices such as motor starter coils, pilot lights, control relays, and solenoid valves. As was true for the discrete input module, the discrete output module also works on a digital, or *ON* and *OFF*, basis. When the processor has made a decision to turn a specific device *ON*, the processor places a 1 in the memory location assigned to that output device, and later in the process, the information is communicated by way of the backplane of the I/O rack to the output module, and the required real-world device will be turned *ON*. Similarly, when the processor determines that a device needs to be turned *OFF*, a 0 is placed in the device's memory location and the device will be turned *OFF*. The output module acts like a remote control switch that is controlled by the processor for turning output devices *ON* and *OFF*.

Output modules are generally classified as AC and DC. Both types offer a wide range of voltages, typical for the input modules discussed earlier. Read the information sheets carefully to determine that the module selected is appropriate for the load (output device) that is to be controlled.

Output modules are sized by the number of output devices that can be connected to them. The number of terminals, or points, for connecting output devices are typically 4, 8, 16 and 32. Output modules have a current rating for each terminal or connection point, as well as an overall rating for the module. The individual terminal rating indicates the maximum current, or load, that can be controlled. This rating is a **continuous duty rating**. A **surge rating** is usually provided, indicating the maximum current that can be controlled and the length of time. The time rating may be given in milliseconds or in cycles. A typical current rating for a 120 V AC output module would be: 1.5 amperes maximum continuous duty, with a surge current rating of 4 amperes for 8.3 msec (1/2 cycle).

The surge current rating is necessary for the inrush, or "pull-in," current that is present when motor starter coils, solenoids, and other inductive loads are initially energized. Once the load has been energized, the "hold-in," or "seal-in," current is much less, and the continuous duty rating of the module is sufficient.

Each module also has a **total current rating**. The total current rating must be determined from the manufacturer's literature, not by simply adding the ampere rating per point. To further clarify this point, consider the rating of one manufacturer's 16-point 120 V AC output module. Each point is rated for 2 amperes continuous duty, yet the maximum current rating for the module is only 8 amperes. Why is the total rating not 2 amperes times 16 points, or 32 amperes? The answer has to do with the way the module dissipates the heat generated by the current flow in the module. Normally, no fans or other external methods of cooling are used, and the heat that is generated within the module is dissipated using **heat sinks**. Heat sinks work on convection alone, and this limits the amount of heat that they can effectively dissipate. The total current rating that a module is given, therefore, is determined by the ability of the module to dissipate heat. Thus, the lower current rating of the 16-point 120 V AC module shows that the module can only satisfactorily dissipate the heat from 8 amperes of continuous current flow, not the full 32 amperes that would be possible if all loads were operating at the maximum 2 amperes rating. In reality, however, how likely is it that all 16 loads connected to the module would be on and operating at their full 2-amperes capacity?

Subjecting the module to higher-than-rated current loads creates excess heat in the module. This excess heat has a detrimental affect on the electronic components and leads to shortened operating life and/or component failure. The ambient temperature that the I/O operates in is another factor that must be considered. PLCs are normally designed to operate in a temperature range of 32°–140°F or 0°–60°C. Operating in temperatures above these limits affects the module's ability to dissipate heat and can lead to early component failure.

Another PLC manufacturer rates their 16-point, 120 V AC output module at .5 amperes per point, but starts to derate the current level when more that 50% of the points are *ON*, and/or when the temperature exceeds 40°C. With each manufacturer rating their output modules differently, it is necessary to carefully review the literature when specifying modules.

AC Output Module

The internal circuitry for 1 point of a typical AC output module is shown in Figure 2–15. The AC output module usually consists of a **Triac** (shown in the figure). However, some manufacturers use a **Silicon-Controlled Rectifier** (**SCR**) instead of a Triac. When the processor determines that the output is to be turned *ON*, a signal is sent across the backplane of the I/O rack and the LED (light emitting diode) is turned *ON*. The light from the LED causes the photo transistor to conduct, which provides current for the gate of the Triac. This portion of the output module optically isolates the logic section of the processor from the line voltage of the output devices.

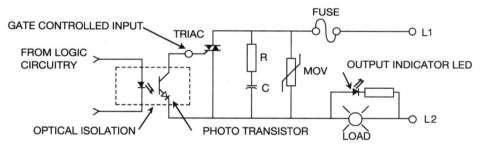

Figure 2–15 Typical AC Output Module Circuit

The Triac is used as an electronic switch to turn output devices *ON* and *OFF*. The Triac itself is the equivalent of two silicon-controlled rectifiers in inverse parallel connection with a common **gate**. The gate controls the switching state (*ON* or *OFF*) of the device. Once a signal is applied and the break-over voltage point is reached on the gate (normally 1 to 3 V), the Triac freely conducts in either direction, completing the path for current flow to the output device.

A Triac is a solid block of crystalline material and is more sensitive to applied voltages and currents than standard relay contacts. Triacs are also limited to a maximum peak applied voltage, and, if this value is exceeded, a "dielectric-type" breakdown can result, causing a permanent short-circuit condition. A snubber circuit that consists of a resistor in series with a capacitor and a metal oxide varistor is used to protect the Triac from damage from electrical noise and voltage spikes. The resistor and capacitor form an RC circuit that is used to control the rate at which voltage builds up in the circuit. If the voltage rises too fast, the capacitor absorbs, and the resistor dissipates, excess voltage. The metal oxide varistor (MOV) is designed to break down, or conduct, at certain voltage levels. In a 120 V AC circuit, the peak voltage is approximately 170 V.

The MOV would typically be set to break down at 190 V. When the MOV conducts, it clips the excess voltage and thus prevents damage to the module.

Triacs constructed of a solid "block of material" have some characteristics that are not found with standard relay contacts. The Triac, rather than having *ON* and *OFF* states, actually has low- and high-resistance levels, respectively. In its *OFF* state (high resistance), a small leakage current still flows through the Triac. This leakage current, which is usually only a few milliamperes or less, normally causes no problem. When low-resistive pilot lights are connected to AC output modules, a faint glow of the filament may be detectable, even when the module is *OFF*. Similarly, the coils of some small control relays or solenoids may produce a detectable hum due to the Triac leakage current, even though the Triac is technically *OFF*. This small leakage current also causes false readings in some digital and analog meters. Troubleshooting techniques for them are covered in Chapter 20.

While Triacs are capable of carrying surge currents higher than their continuous current ratings, such surges must be of short duration (1/2 to 1 cycle) and not repetitive. Exceeding the manufacturer's listed surge values or the maximum continuous current rating, usually referred to as maximum RMS on-state current, results in a permanent short circuit.

After an output module is installed in the I/O rack, the actual real-world output devices are connected. Figure 2–16 shows the proper termination for an 8-circuit AC output module. Each output device has two wires; one wire from each output device is wired to the L2 potential. The other wire from each device is wired to one terminal of the output module, as shown in the figure. L1 is then connected to a common terminal on the module to supply the other potential for the output devices.

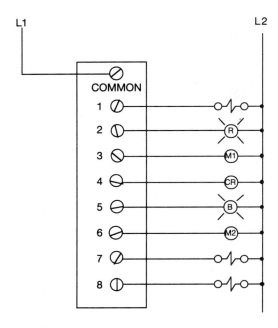

Figure 2–16 Field Wiring for AC Output Module

Figure 2–17 shows two simplified circuits, each including an output device. In the first circuit, the output is wired to the L2 potential and the single-pole switch is used to control the L1 potential to complete the circuit. In the PLC circuit, the output is again wired to L2, but the switch is replaced by the output module which is wired to the L1 potential. Simply stated, the output module may be viewed as an electronic switch that is controlled by the PLC's processor.

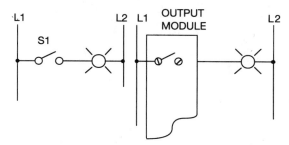

Figure 2–17 Traditional Wiring for an Output Device (A), Compared to PLC Wiring (B)

Output Fuses

In order to prevent damage to output modules, it is important not to exceed the current rating or to exceed its inrush capability. It is also important that output modules be protected from short circuits and ground faults. To provide protection for overcurrent, short circuits and ground faults, output modules are always fused. The number of fuses used vary with each manufacturer. Some modules have an individual fuse for each output terminal, or point, others have one fuse for each 8 outputs, while still others use one fuse to protect all 16 points on their output module. Some PLCs, like the Allen-Bradley SLC 500, do not come with internal fuses, and fuses must be added to protect the outputs.

Most PLCs come with "blown-fuse" indicators to show that a fuse has blown. Some modules have a "blown-fuse" indicator for each output terminal. If a fuse is blown on output terminal 6, an indicator lamp at terminal 6 will be lit to indicate the location of the blown fuse. Other modules have only one "blown-fuse" indicator for the whole module. This one indicator lamp only indicates that a fuse has blown; it does not indicate its location, and it is up to the electrician or technician to determine which fuse (or fuses) is faulty. When individual indicators are used for each output terminal, troubleshooting is greatly simplified. When only one indicator is used for the entire module, however, a blown-fuse indicator merely tells you that a fuse has blown, but does not indicate which one. Figure 2–18 shows a typical AC output module with the fuses and blown-fuse indicators identified.

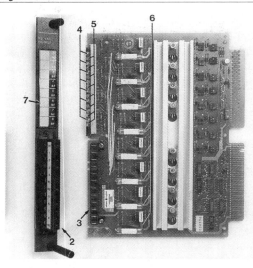

Figure 2–18 AC Output Module with Fuses and Blown Fuse

As you may expect, access to the fuses for changing blown fuses varies with the manufacturer. Some modules must be removed from the I/O rack and a cover removed before the fuses can be changed, while other modules provide direct access to the fuses on the front edge of the module. For modules that must be removed to change fuses, electricians or technicians must be sure to turn the power *OFF* before removing or reinserting the module.

It is important that when a blown fuse is removed, its replacement fuse be of the same voltage rating, amperage rating, interrupting rating, response time, and physical size, as recommended by the manufacturer.

To speed troubleshooting and fuse replacement, many plants now add additional fuses to each output circuit. Terminal blocks are available that have built-in fuse holders and individual blown-fuse indicators. A separate fuse can then be wired in series with each output device, external to the output module. The fused terminal blocks are then mounted in a convenient location for easy access and troubleshooting.

Status Lights
Status lights are provided for each output point to indicate when that point has been turned *ON*. For troubleshooting purposes, it is important to understand how these status lights are wired. If the power for the lights comes from the processor side of the module, an illuminated light indicates that the processor has sent a signal to turn *ON* the output attached to *that* particular point. It is *not* an indication that current is flowing to the output device. If the status light is powered from the actual output power, then it is an indication that current is flowing to the output device.

Some modules have two status lights. One is referred to as the **logic light**, and the other as the **output light**. When the logic lamp is lit, it indicates that the logic to turn on the output device has been sent from the processor. When the output light is lit, it indicates that there is a path for current flow to the output device.

The status lights are a powerful troubleshooting tool, but it is important to understand exactly what the status lights indicate, and not to read more information into them than they are able to provide.

Module Keying

Figure 2–19 shows an Allen-Bradley 240 V AC output module. This AC output module looks just like the AC or DC input module discussed earlier (Figure 2–10c). This is not only true for Allen-Bradley modules, but also for other manufacturer's products as well. In all instances, the manufacturer's label and/or color codes the fronts of all modules to distinguish between the different types (AC input, DC output, TTL, Analog, and so on). Most manufacturers have designed each module so it can be **keyed**. In Figure 2–19, item 6, the module has been notched in two places. Installing keying bands on the Allen-Bradley I/O rack backplane connector (Figure 2–20) where a specific module is to be installed prevents any module, other than the type for which the connector is keyed, from being installed in that connector or slot. Figure 2–21 shows a close-up view of what the keying band looks like and how it is installed. Each type of module has a unique combination of notches. This feature prevents inadvertent or accidental replacement of the wrong type of module, say an input module, into a slot that is already wired to output devices. To prevent damage and downtime, it is important that the keying system be used.

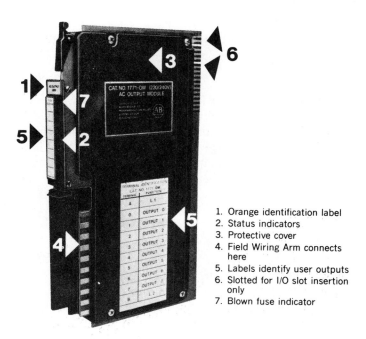

1. Orange identification label
2. Status indicators
3. Protective cover
4. Field Wiring Arm connects here
5. Labels identify user outputs
6. Slotted for I/O slot insertion only
7. Blown fuse indicator

Figure 2–19 Notched AC Output Module
(Courtesy of Allen-Bradley)

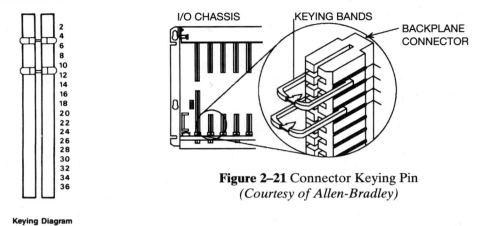

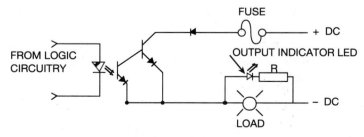

Keying Diagram

Figure 2–20 Backplane Connector Keying
Diagram
(Courtesy of Allen-Bradley)

Figure 2–21 Connector Keying Pin
(Courtesy of Allen-Bradley)

There are some manufacturers that do not use the physical keying system, but rely on software configuration to prevent inadvertent or accidental replacement of a wrong module. Each module is given a unique "signature" that is relayed to the processor. If the signature does not match the software configuration, a fault will occur and prevent the PLC from operating.

DC Output Modules

DC output modules are basically the same in operation as AC output modules. The difference is the use of a power transistor instead of a Triac for the control of output current. The power transistor has a quicker switching capability than the Triac; therefore, the response time for DC modules is faster than for AC modules. The RC circuit and MOV used in the AC output module is replaced with a diode to provide protection from electrical noise and spikes. A typical DC output circuit is shown in Figure 2–22.

Figure 2–22 Simplified DC Output Module Circuit

DC output modules are available in ranges from 12–240 V DC depending on the manufacturer. It is important to check to make sure that the module is appropriate for the voltage and current level of the output device that is intended for connection to the module. DC modules will be fused like their AC counterparts and also have blown-fuse indicators, as well as status lights. Figure 2–23 shows how the output devices are wired to a DC output module.

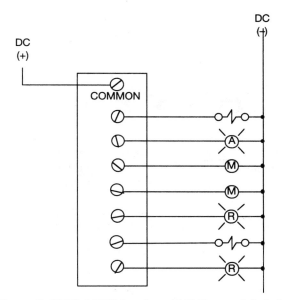

Figure 2–23 Field Wiring for a DC Output Module

Sourcing and Sinking

The terms **sourcing** and **sinking** refer to the manner in which DC devices are wired. To properly interface DC devices with the outside world, the difference between sourcing and sinking must be understood.

Figure 2–24 is an example of a sourcing application. The positive potential is connected to the input module and the negative potential is connected to the input device. Using conventional current flow (+ to −), it is said that the input module is the source of supply for the real-world input device. Stated simply, the input device receives current from the input module. In electronics, if a device (input module) provides current, or is the source of current, it is said to be sourcing.

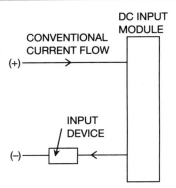

Figure 2–24 Input Module
Sourcing Application

Figure 2–25 is an example of a sinking application. In this configuration, the positive potential is connected to the input device and the negative potential is connected to the input module. In this case, using a conventional current flow of + to –, the input device is said to be providing current to the input module. If the device (input module) is receiving current, it is said to be sinking.

When an output module is connected positive, as shown in Figure 2–26, and provides current to the outside world (output device), the module is referred to as sourcing.

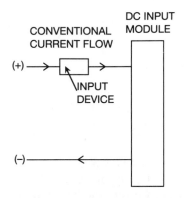

Figure 2–25 Input Module
Sinking Application

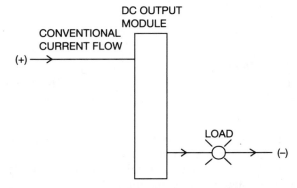

Figure 2–26 Output Module Source Capability

Figure 2–27 shows an output module wired negative, and the output device wired positive. When the output device (outside world) provides the current to the module, the module is referred to as sinking.

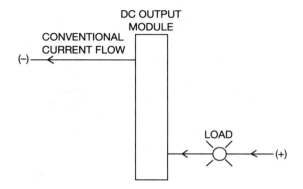

Figure 2–27 Output Module Sink Capability

There is always a great deal of confusion surrounding sourcing and sinking as it applies to input and output modules and devices. As a general rule, sinking modules are used with output modules when interfacing with electronic equipment (TTL or CMOS compatible), while sourcing modules are used for such DC loads as solenoids. The safest way to make sure that connections to DC devices are correct is to ignore the terms sink and source, and select a module that works for the application, based on the manufacturer's specifications and wiring diagram.

Contact Output Modules

Contact modules have electromechanical relays mounted on a printed circuit board that is inserted, or plugged into, the I/O rack. A signal from the processor energizes the coil of the electromechanical relay which, in turn, opens or closes a set of contacts. Each set of contacts is isolated and can be ordered normally open (N.O.) or normally closed (N.C.). This type of module is used when extra current ratings are required or when it is desirable to isolate loads of different voltages, or phases, from the same source. A contact output module is also used when the leakage current of a standard AC output module would affect the control process.

For Example, a contact module can be used with a Variable Speed Drive application. Assume that the drive has been programmed for multiple speed settings. To select a given speed, a circuit must be completed at two points on a terminal strip mounted on the drive. One set of contacts is used for each speed setting. As the drive supplies its own power, all that is needed is to switch the control power through the contacts on the contact module. As the contacts are isolated, there is no possibility that the power from the drive system can damage the PLC.

Interposing Relay

When it is necessary to control loads larger than the rating of an individual output circuit, a standard control relay, which has a small inrush and sealed current value, is connected to the output module. The contacts of the control relay, which are generally rated at 10 amps, can then be used to control a larger load. This method of control is a common practice for NEMA size 4 and large motor starters, depending on the rating of the output module. When a control relay is used in this manner, it is called an **interposing relay** (Figure 2–28).

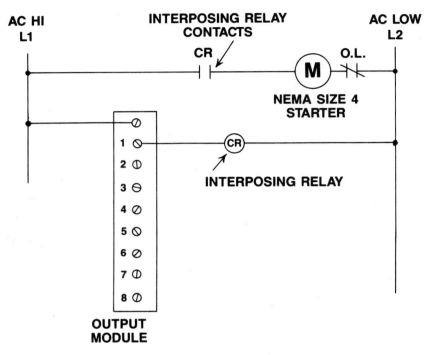

Figure 2–28 Interposing Relay

Reed Relay Output Module

The reed relay type output module is used when dry reed relays are desirable. They may be used for low-level switching (small current–low voltage), multiplexing analog signals, or for interfacing controls with different voltage levels. The voltage range of the reed relay contacts are normally in the range of 0–130 V AC or 0–125 V DC with a current rating of 1 ampere. Reed relay output modules are cheaper than normal solid-state AC/DC output modules and are usually recommended when indicator lamps are the output devices. Reed relay modules are available with normally open (N.O.) contacts, normally closed (N.C.) contacts, or a combination of both N.O. and N.C. contacts, again depending on the manufacturer.

Transistor-Transistor Logic (TTL) I/O Modules

TTL input modules are designed to be compatible with other solid-state controls, sensing instruments, many types of photoelectric sensors, and some 5 V DC level control devices. TTL output modules are used for interfacing with discrete or integrated circuit (IC) TTL devices, LED displays, and various other 5 V DC devices.

Analog I/O Modules

Analog input modules are used to convert analog signals from analog devices that sense such variables as temperature, light intensity, speed, pressure, and position to 12-Bit Binary or to 3-digit Binary-Coded Decimal (BCD), depending on the manufacturer, for use by the processor. The conversion from analog to digital is accomplished with an Analog-to-Digital Converter, or ADC. The analog output module changes the 12-Bit Binary or 3-digit BCD value used by the processor into analog signals using a Digital-to-Analog Converter, or DAC. These analog signals can be used for speed controllers, signal amplifiers, or valve positioners. Binary and Binary-Coded Decimal (BCD) is covered in Chapter 6.

NOTE: PLC manufacturers are introducing new and special application modules almost daily. A few modules have been discussed in this chapter for a basic understanding only. The local PLC representative(s) should be contacted for a full and complete list(s) of modules that are available.

Safety Circuit

The National Electrical Manufacturing Association (NEMA) standards for programmable controllers recommends that consideration be given to the use of emergency-stop functions that are independent of the programmable controller. The standard reads in part: "When the operator is exposed to the machinery, such as loading or unloading a machine tool, or where the machine cycles automatically, consideration should be given to the use of an electromechanical override or other redundant means, independent of the controller, for starting or interrupting the cycle."

While programmable logic controllers of today are rugged and dependable, where safety is concerned *do not* depend on the solid-state devices and circuitry of the PLC, or PLC program. The NEMA recommendation recognizes the importance of a hard-wired emergency-stop circuit, or E-Stop, (shown in Figure 2–29) to remove power to the output devices. A second set of MCR contacts could be added in the X1 line to remove power to the input devices. It is common, however, to put the MCR contacts on the output side only so that the inputs can remain energized for troubleshooting.

It is also worth noting that solid-state output devices usually (though not always) fail shorted, rather than in an *open* condition. By failing in a shorted or *ON* condition, an added safety hazard is possible if a hard-wire emergency stop (E-Stop) or master control is not included as part of the PLC installation.

Rack Installation

Before installing a rack or chassis, consideration must be given to the following:
- temperature
- dust
- vibration
- humidity
- field wiring distances
- troubleshooting accessibility

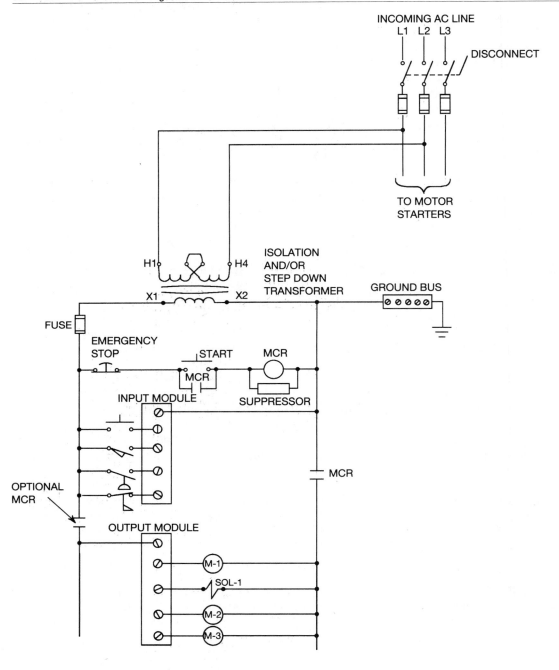

Figure 2–29 Power Distribution with Master Control Relay (MCR) for a Grounded AC System

The ambient temperature of the proposed location should not be lower than 32°F or higher than 140°F (0°C and 60°C). Fans are normally not used with I/O racks, and all cooling of the electrical or electronic components is accomplished by convection. Convection cooling is accomplished when warm air caused by heat in the components rises and creates a movement of air. This movement of air draws cool air in through the bottom of the rack and expels warm air out through the top. To maintain efficient convection cooling, it is important that the rack be installed correctly and not used as a shelf for notebooks or other material that would impede or block the natural flow of heat up through the rack.

During initial installation it is common practice to cover the top of the rack to prevent any scrap wire, stripped insulation, screws, and nuts, from falling into the I/O modules, power supply, or processor which could cause a short circuit or other electrical failure. The protective cover must be removed after installation to assure that proper cooling can take place.

Under adverse conditions, when the ambient temperatures exceed the manufacturers' recommended maximums, rack or chassis fans can be used. The fans are mounted under the I/O rack (Figure 2–30), and forces air up through the I/O assembly to aid with the movement of air. The addition of the fans reduce air temperatures within the rack and substantially increases reliability.

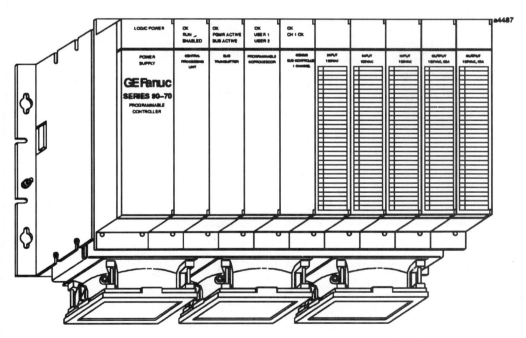

Figure 2–30 Rack Fans Mounted Under Rack to Aid Cooling

When the temperature is expected to go below 32°F, a thermostatically-controlled heater is used inside the enclosure to prevent condensation.

Dust can also cause a problem in the I/O rack when it accumulates on the electronic components of the modules, power supply, or processor. Accumulated dust prevents the components from dissipating heat effectively. A dust-tight enclosure with a cover and gasket can be used to prevent problems that too much dust can create. It is important to remember that any enclosure used to house PLC components must be large enough to allow for proper air circulation and heat dissipation. If the enclosure is too small, the heat will build up inside the enclosure and have a detrimental effect on the electrical or electronic components. The installation manual usually specifies the minimum size of enclosure that can be used.

Excessive vibration can also lead to early component failure. It is important to mount PLC equipment on solid, nonvibrating surfaces. Vibration affects from equipment must be minimized to assure proper longevity for the equipment.

Humidity, while normally not a problem, must be considered when installing a PLC. Allen-Bradley, for example, rates their PLCs for operation in a humidity range of 5%–95% (without condensation). Some manufacturing processes, however, create high humidity (high moisture content) conditions. Exposing electronic equipment to extremely high humidity environments over an extended period of time can reduce component life and affect operation. Evaluate the environment carefully and mount equipment in an area that minimizes the exposure to high humidity and moisture.

While it is important that the controller and programming unit be installed or mounted in a control center or other central location, the use of remote I/O racks allows the input and output modules to be installed close to the actual operating equipment (Figure 2–31).

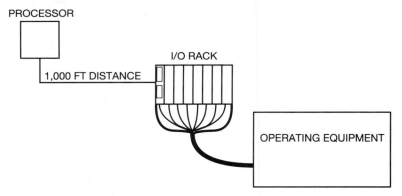

Figure 2–31 Remote I/O Rack Close to Operating Equipment

By mounting the I/O rack close to the actual equipment, the amount of conduit, cable, and other associated wiring and labor costs will be decreased. The only wiring needed for communication back to the processor will be a shielded-twisted pair, twin axial cable, fiber optic cable, and the like. By having the input and output modules located close to the process or driven equipment, troubleshooting is also easier and more efficient. As discussed earlier in this chapter, each input and output module has status lights that indicate whether an input or output device is *ON* or *OFF*. Having this capability close to the actual equipment shortens troubleshooting time and increases production.

Before mounting the racks and other associated equipment, give careful consideration to location and accessibility. If access is restricted or difficult to reach, troubleshooting, repair, and maintenance is more difficult and time consuming.

Electrical Noise (Surge suppression)

Electrical noise is generated whenever inductive loads such as relays, solenoids, motor starters, and motors are operated by "hard contacts" such as push buttons, selector switches, and relay contacts. The noise, or high transient voltages (spikes), are caused by the collapsing magnetic field when the inductive device is switched OFF. The level of the voltage spike can be very high and is capable of causing erratic operation of the processor and/or output module, or can cause permanent damage to the module. The interference caused by these voltage spikes and the accompanying electrical noise is often called **Electromagnetic Interference** (**EMI**). There are several steps that can be taken to reduce or eliminate the effects of EMI. Two of the most common are **isolation** and **suppression**.

Isolation of the electrical noise is accomplished by installing an isolation transformer for the PLC system (Figure 2–32) to supply the power for the controller and the input circuits. The figure shows a constant voltage transformer, but a standard step-down transformer of the proper size also effectively isolates electrical noise.

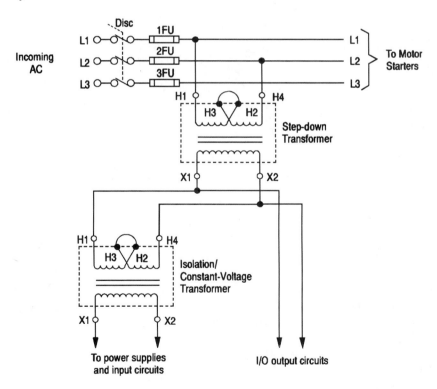

Figure 2–32 Reducing Electrical Noise with an Isolation Transformer

A second method in reducing EMI is to install surge suppression networks or devices on the individual motor starters, motors, and solenoids. These suppression devices can consist of an RC circuit (resistor/capacitor), a Metal Oxide Varistor (MOV) or a MOV and RC combination for AC loads and a diode for DC coils. The collapsing magnetic field of the inductive device is, in a sense, dissipated by the suppression network and reduces the effects of EMI. Figure 2–33 shows typical installations of the various types of suppression devices.

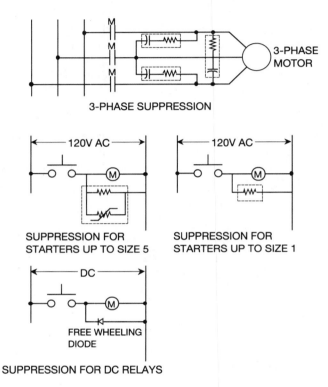

Figure 2–33 Typical Equipment Grounding Configuration

The type of surge suppressor to use depends on the size and type of load. An equipment representative or local electrical distributor should be consulted for help with selection and application.

Grounding

With solid-state control systems, proper grounding helps eliminate the effects of electromagnetic induction. Figure 2–34 shows a typical installation using an equipment grounding conductor to connect several PLCs and/or I/O racks together. The equipment grounding conductor is attached to the metal frame of the PLC and/or I/O rack with a ground lug. A detail of the connection is shown in Figure 2–35.

NOTE: Check local codes and manufacturers' specifications to ensure proper installation.

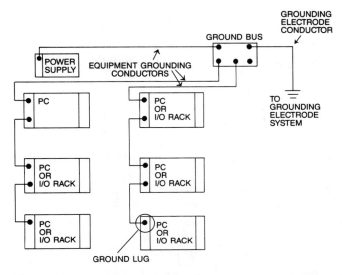

Figure 2–34 Typical Equipment Grounding Configuration

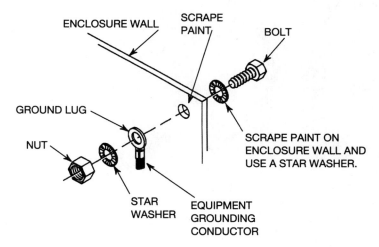

Figure 2–35 Detail of Grounding Lug Attachment to PLC

I/O Shielding

Certain I/O modules such as TTL, analog, and thermocouples require shielded cable to reduce the effects of electrical noise. The cable shield, which surrounds the cable conductors, shields the conductors from electrical noise.

When installing shielded cable, it is important that the shield only be grounded at one end. If the shield is grounded at both ends, a ground loop is created and can introduce ground currents that may result in faulty operation of the processor.

As a properly grounded I/O rack is already connected to earth ground through an equipment grounding conductor, the shield should be terminated at the I/O rack, *not* at the device end.

Figure 2–36 shows the shield of a shielded cable connected to the I/O rack frame.

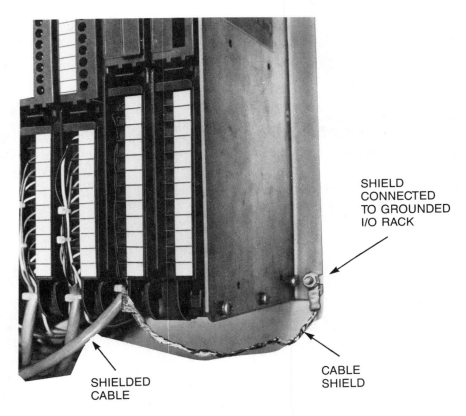

Figure 2–36 Cable Shield Connected to Grounded I/O Rack
(Courtesy of Allen-Bradley)

Figure 2–37 shows a shielded-twisted pair cable connected to a sensing device and I/O rack. Note that the shield is connected at one end only.

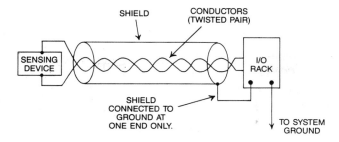

Figure 2–37 Shields Connected to Ground at One End Only

Additional methods of noise reduction are as follows:
- Mount equipment in steel metal enclosures, when possible, because metal helps protect against electromagnetic interference (EMI)
- Separate I/O and PLC wiring from the motor and other large loads to reduce the possibility of induction in the control circuits. This is usually accomplished by installing the control wiring in one raceway, or cable-tray, and the power circuits in another raceway or cable-tray with physical distance or separation between the two

Chapter Summary

The I/O rack houses the individual input and output modules that are connected to real-world devices. The input modules act as an interface between the actual input devices and the processor, while the output modules act as an interface between the actual output devices and the processor. The status (*ON* or *OFF*) of the input devices is communicated to the processor; the processor makes a decision, and in turn communicates to the output modules to turn *ON* or *OFF* the output devices that are connected to the output module. The processor may be connected to the I/O rack by way of an interconnecting cable(s), through a bus duct, or it may be mounted in the same rack as the I/O.

The I/O section is divided into two categories: fixed and modular. Discrete I/O modules operate on digital, or *ON* and *OFF* signals, whereas analog I/Os operate on a variety of signal levels and types. Input and output modules are available in a variety of voltages and normally can control 4, 8, 16 or 32 individual input or output devices. Optical coupling, or isolation, is used to protect the low-voltage (5 V DC) side of the processor from the line-voltage input and output signals that can be as high as 240 V.

AC output modules typically use Triacs for switching *ON* and *OFF* the actual output devices. Triacs when in the high-resistive, or *OFF* state, have a small leakage current that flows through the output device. When Triacs fail, they normally fail in the *ON* condition. Fuses used for protection of output modules are carefully selected by the manufacturer for current and time characteristics, and only fuses recommended by the manufacturer should be used to prevent possible damage to the equipment.

PLC troubleshooting is simplified by the addition of status lights on the I/O modules. The lights indicate which inputs are *ON* or *OFF* and which outputs are *ON* or *OFF*. Indicator lights also indicate if an output module has a blown fuse. To prevent an incorrect module from being installed in a given rack slot, the modules are often keyed.

There is a wide variety of input and output modules available that fit almost any application. Care must be taken to ensure that the module has the correct voltage, current, and time characteristics. The various PLC manufacturers continue to introduce innovative new modules to meet the changing requirements of automated equipment and process control.

Proper installation of PLC equipment requires that the environment (dust, heat, humidity, and vibration) be considered, as well as the physical location for access and troubleshooting. Reduction and/or elimination of electrical noise, voltage spikes, voltage variation, and the like, is necessary to ensure proper operation of the system.

REVIEW QUESTIONS_____

1. Describe briefly the purpose of the I/O section.
2. State two reasons for employing optical isolation.
3. Draw an input module with four input devices, show all necessary electrical connections, and identify potentials L1 and L2.
4. Draw an output module with four output devices, show all necessary electrical connections, and identify potentials L1 and L2.
5. T   F   Triacs are susceptible to "dielectric-type" breakdown if the maximum peak voltage level is exceeded.
6. Briefly describe why a hard-wired emergency-stop circuit is recommended for PLC installations.
7. Briefly describe the function of an interposing relay.
8. T   F   I/O modules are keyed to prevent unauthorized personnel from removing them from the I/O rack.
9. Which of the following are *not* normally sources of electrical noise?
   a. solenoid
   b. relay
   c. indicator lamp
   d. motor starter
   e. motor
   f. overload heaters
10. T   F   To ensure maximum benefit of shielding, the shield of a shielded cable must be terminated and grounded at both ends.

11. E-Stop refers to
    a. extra stop
    b. emergency stop
    c. every stop
    d. elevator stop
    e. energy stop
12. T   F   Electromagnetic interference (EMI) can be reduced with the proper grounding of equipment.
13. Solid-state output devices tend to
    a. never fail
    b. fail in the *open* or *OFF* condition
    c. fail in the *shorted* or *ON* condition
    d. not be affected by overload
14. List three environmental considerations when installing PLC equipment.
15. What type of tool or object should be used to change the position of DIP switches?

Chapter 3 _____ Processor Unit

Objectives
After completing this chapter, you should have the knowledge to
- Describe the function of the processor.
- Describe a typical program scan.
- Identify the two distinct types of memory.
- Describe the function of the *watchdog timer*.
- Identify various memory designs.
- Identify and explain the two broad categories of memory use.
- Identify different peripheral devices that are used with a programmable controller.

The processor unit houses the micro-processor, memory module(s), and the communications circuitry necessary for the processor to operate and communicate with the I/O and other peripheral equipment. The DC power required for the processor is provided either by a power supply that is an integral part of the processor unit, or by a separate power supply unit, depending on the manufacturer. The processor, or "brain," of the programmable logic controller is the decision-maker that controls the operation of the equipment to which it is connected. The processor controls the operation of the output devices that are connected to the output modules based on the status of the input devices and the program that has been entered into memory (Figure 3–1).

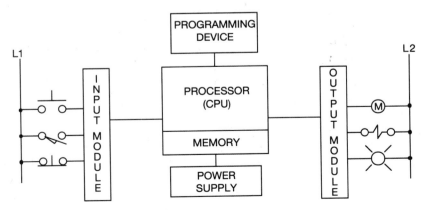

Figure 3–1 Basic PLC Configuration

Processors are available that control as few as 8, or as many as 40,000, real-world inputs and/or outputs. The size of the processor unit to be used is dependent on the size of the process(es) or driven equipment to be controlled. The larger the number of input and output devices that are required for the process, the more powerful the processor must be to properly control the number of I/O that will be connected. One PLC can control more than one machine or process line and is limited only by the I/O required, physical distance, and memory capacity of the PLC used.

NOTE: It is difficult to discuss processor unit configuration due to the differences in PLC hardware from the various manufacturers. The discussion that follows is general and is not intended to cover all the PLCs on the market today. It should also be noted that when pictures are used to illustrate a given configuration or concept, one of a particular manufacturer's models is illustrated; however, the manufacturer also has other models larger and/or smaller, and in different configurations.

The Processor

The processor may be a self-contained unit, or may be modular in design, and plug directly into the I/O rack as shown in Figure 3–2. Whatever the configuration, the processor consists of the micro-processor, memory chips, circuits necessary to store and retrieve information from the memory, and communication circuits required for the processor to interface with the programmer, printer, and other peripheral devices. The memory and communication circuits can be modules separate from the micro-processor module. The actual hardware configuration will depend upon the PLC.

Figure 3–2 Allen-Bradley SLC 500 with a SLC 5/01 Processor (CPU) Installed
(Courtesy of Allen-Bradley)

The **micro-processor** is the device that
1. Monitors the state or status (*ON* or *OFF*) of the input devices.
2. Systematically solves the logic of the user program.
3. Controls the state of the output devices (*ON* or *OFF*).

This three-step process is referred as the processor **scan**.

When the PLC is powered up or turned *ON*, the processor runs an internal self-diagnostic, or self-check, prior to initiating its first scan. If any part of the processor system is not functioning, such as a faulty memory, improper communication with the I/O section, or failure in a remote rack, the processor fault light or other indicator light comes *ON*. With some systems, if a monitor is connected, a written explanation or fault code will appear on the VDT screen. Some systems use status words to indicate the hardware or software that has malfunctioned. Status words can be included in the program so that when a malfunction is detected, an alarm will sound to alert the operator that there is a problem.

Once the processor has passed the self-diagnostic check, it is ready to go to work. Figure 3–3 illustrates a typical PLC scan. In the first step of the scan, the processor determines the status of the input devices. It does so by looking at the memory locations that have been designated for all of the input devices. Remember, as stated earlier in the text, the actual status (*ON* or *OFF*) of any input device is stored in a memory location as either a 1 or a 0. A 1 indicates that a device is *ON* or closed, while a 0 indicates that the input device is *OFF* or open. Based on the 1s and 0s, the processor determines the actual condition of all of the input devices.

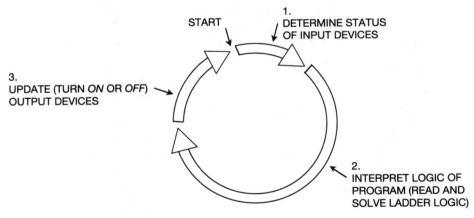

Figure 3–3 Typical Processor Scan

The second step in the processor scan is to interpret the logic of the program that has been written and stored in the processor memory. Based on the program requirement, the processor will turn the required output devices *ON* or *OFF*, which is the third step in the processor scan. This third step is referred to as **updating** the outputs. This updating process occurs once during each scan.

The scan is continuous and the three-step process is repeated over and over every few milliseconds. To summarize, the three steps of the scan are
1. Determine the status of the input devices (*ON* or *OFF*).
2. Read and solve the logic of the program (ladder logic).
3. Update the output devices (turn *ON* or *OFF*).

The time it takes to complete one scan will vary from a fraction of a millisecond to 100 + milliseconds, depending on the size of the program. Each manufacturer lists the scan time based on the use of 1K of program memory. A typical scan time is 3–5 msec per 1K (1024 words) of memory used. Scan time is also affected by factors other than just the amount of memory that was used in the program. If a program that is written to control a particular piece of equipment uses 1K of memory with a 5 msec scan time, the program is scanned approximately 200 times each second. Increasing the size or complexity of the circuit uses additional memory, and extra words (memory) increases the scan time. A program with 3K of memory takes approximately 15 msec for each scan.

The use of remote I/O racks also increases the scan time. The added time is required because the status of the input devices must be transmitted from the remote I/O location to the processor location.

NOTE: As part of the processor's internal self-diagnostic system, a **watchdog timer** is used. The watchdog timer is preset to an amount of time that is slightly longer than the scan time would be under normal conditions. At the start of each scan, the watchdog timer is turned *ON* and starts to accumulate time. If the program is correct, the program scan will be completed prior to the time set on the watchdog timer, and at the end of each scan, the watchdog timer is reset to 0. If for some reason the program scan is not completed in the allotted time, indicating that there is a problem with the program, the watchdog timer will time out, which puts the processor into a faulted condition. The range of the timer is software selectable (adjustable) on many PLCs.

Normally, before any output devices can be turned *ON* or *OFF*, the processor has to scan the entire program that is in user memory. The program may only be a few Rungs long, or it may be hundreds of pages in length, depending on the equipment that is being controlled. Some input devices operate so fast that the entire program is read and solved before the outputs are updated, and the input device may have changed positions more than once since the processor originally determined its status at the start of the scan. The same may be true for an output device that needs to be updated sooner than a regular scan will allow. To solve this problem, many PLCs have special program instructions that allow critical or high-speed input and output devices to be updated sooner than would be possible under normal scan conditions. The special instructions actually interrupt the scan when it is reading the program and allows I/O devices to be updated immediately.

Scan time of a given PLC and its ability to provide special function programming to accommodate high-speed or critical input and output devices must be a consideration when initially selecting a PLC.

NOTE: Additional information, and a more detailed discussion on how the processor scans the user program, is covered in Chapter 10.

The memory section of the processor consists of hundreds or thousands of locations where information is stored. In the broadest sense, the memory is divided into two classifications: user and storage. The **user memory** is for the storage of the user program which contains the relay logic, or instructions, that control the driven equipment or the process. The **storage memory** is used to store information such as input/output status, timer or counter preset and accumulated values, and internal control relays, etc. that are necessary for the processor to control the equipment or process. The actual memory structure of various PLC manufacturers will be covered in Chapter 5, while the purpose or use of each memory is covered later in this chapter.

Memory chips used in the processor can be separated into two distinct groups: **volatile** and **nonvolatile**. A volatile memory is one that loses its stored information when power is removed. Even momentary loss of power erases any information stored or programmed on a volatile memory chip.

A nonvolatile memory has the ability to retain stored information when power is removed, accidentally or intentionally.

To protect a volatile memory, backup batteries are included in the processor power supply. The batteries may be "D" size dry cells, rechargeable nickel cadmium, lead acid, or nonrechargeable alkaline or lithium types.

> ***CAUTION:**   Extra care must be exercised when disposing of batteries, as they are now classified as hazardous waste. Special care must be taken with lithium batteries because they may explode when exposed to, or dropped into, water.

When batteries are included, they may be located in the processor, or in the power supply, depending on the PLC. Wherever they are located, there is a battery indicator light(s) to indicate the condition, or state of charge, of the batteries. Common indicator lights are *BAT OK* and *BAT LOW*. A simpler system uses one light to indicate that the battery condition is normal. When the light goes out, it is a warning that the batteries need to be replaced.

When the battery indicator light comes on, indicating that the batteries need to be replaced, the memory is still protected for a minimum of two weeks. Depending on the size of the memory and the type of batteries used, in many cases the memory remains protected for one year or more with fully-charged batteries. In reality, rarely is the power interrupted or off for more than a few hours.

The type of batteries used and the number required will vary with each manufacturer. Because alkaline and lithium batteries are not rechargeable and must be replaced periodically, care must be taken to always replace the batteries with the type specified, paying special attention to the orientation of each battery in the battery holder to ensure that proper polarity is maintained. Some batteries are available with leads that simply plug into a connector on the PLC. Figure 3–4 shows a drawing of a lithium battery with leads that is mounted on the back of the faceplate cover, and is used to back up the RAM of the GE Fanuc Series 90-20 PLC.

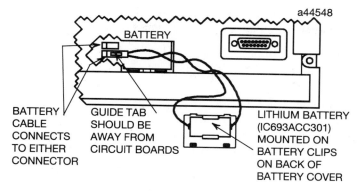

a44548

Figure 3–4 Lithium Battery with Leads
(Courtesy of GE-Fanuc Automation)

*CAUTION: As a general rule, a copy is made of the current program prior to changing the batteries. This copy is referred to as a "back-up" copy, and is used to replace the original program if for some reason the program in memory is lost. The batteries in several PLCs can be changed without turning off the main power. The processor unit of the GE Fanuc Series 90 PLC, shown in Figure 3–5, uses a single lithium battery that protects the volatile memory. Note that a second battery connector has been added and wired in parallel. The new battery is attached to the second connector in order to protect the memory before the old battery is removed. Changing batteries is one of the few maintenance requirements of a PLC. Failure to change the batteries in a timely manner may have serious consequences if a back-up copy of the program is not made. Common sense dictates that a back-up copy be made of every PLC program.

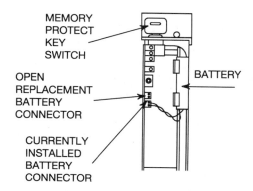

Figure 3–5 Parallel Battery Connection
Used with GE Fanuc 90-70 CPU
(Courtesy of GE-Fanuc Automation)

Memory Types

No attempt will be made to explain solid-state memory types in more than a generalized way for basic understanding. Detailed explanations of solid-state memory types are available in the electronics section of most libraries.

The most common type of volatile memory is **Random Access Memory (RAM)**. Information can be written into, or read from, a RAM chip, and it is often referred to as read/write memory. Information stored in memory can be retrieved or read, while write indicates that the user can program or write information into the memory. Random access refers to the ability of any location (address) in the memory to be accessed or used. RAM is used for both the user memory and storage memory in many PLCs. Since RAM is volatile, it must have battery back-up to retain or protect the stored program. Various forms of RAM include MOS, HMOS, and CMOS-RAM (Complimentary Metal Oxide Semiconductor), one of the most popular, to name just a few.

CMOS-RAM is popular because it has a very low current drain when not being accessed (15 μ amperes), and the information stored in memory can be retained by as little as 2 V DC. A typical fully-charged lithium battery is rated 2.95 V at 1.75 amperes/hours and normally holds or protects a program for 60 days or longer.

Nonvolatile memories are memories that retain their information or program when power is lost, and do not require battery backup. A common type of nonvolatile memory is **Read Only Memory (ROM)**. Read only indicates that the information stored in memory can be read only, and cannot be changed. Information in ROM is placed there by the manufacturer for the internal use and operation of the PLC, and the manufacturer does not want the information changed or altered. PLCs, like other computer-based systems, undergo constant change. When changes are made in the way a system operates, or when new features are added, ROM chips can be replaced to upgrade the PLC.

Other types of nonvolatile memory are PROM, UVPROM, EPROM, EAROM, and EEPROM.

PROM   Programmable Read Only Memory allows initial and/or additional information to be written into the chip. PROM may be written into only once after being received from the PLC manufacturer, and programming is accomplished by pulses of current. The current melts fusible links in the device, preventing it from being reprogrammed. This type of memory is used to prevent unauthorized program changes.

NOTE: Regardless of the memory type, the memory can also be protected by a key switch located on the front of the processor, or on the programming device. With the programmer "locked-out," the program in the processor can be run but not changed. The key switch can also be used to lock the processor out completely and prevent it from running the program.

Another popular method of restricting access to the program is to use "passwords." Passwords restrict access to the program to only those personnel who know the correct password and how to enter it using the programming device. Passwords are often referred to as "software" locks, whereas key switches are referred to as "hardware" locks.

UVPROM–EPROM **U**ltra **V**iolet **P**rogrammable **R**ead **O**nly **M**emory is ideally suited when program storage is to be semipermanent, or additional security is needed to prevent unauthorized program changes. The UVPROM chip is also referred to as EPROM (**E**rasable **P**rogrammable **R**ead **O**nly **M**emory) (Figure 3–6). The EPROM chip has a quartz window over a silicon material that contains the electronic integrated circuits. This window is normally covered by an opaque material, but when the opaque material is removed and the circuitry exposed to ultraviolet light, the memory content can be erased. Once erased, the EPROM chip can be reprogrammed, using a special programmer. After programming, the chip window must once again be covered with an opaque material, such as electrician's tape, to avoid undesirable alteration of the memory.

Figure 3–6 Typical UVPROM or EPROM Memory Chip
(Courtesy of Allen-Bradley)

*CAUTION: Special care and handling of the UVPROM, or for that matter any integrated circuit (IC) chip, must be exercised to ensure that the pins do not become dirty, bent, or subjected to any static electric charges.

EAROM **E**lectrically **A**lterable **R**ead **O**nly **M**emory chips can have the stored program erased electrically. This is accomplished by applying different values of positive (+) and negative (−) voltage values to specific circuit points. When erasing an EAROM chip, the voltage values and pin locations are supplied in the manufacturer's literature. Once erased, the EAROM chip can be reprogrammed.

EEPROM **E**lectrically **E**rasable **P**rogrammable **R**ead **O**nly **M**emory is also referred to as Double EPROM and E^2PROM. EEPROM is a chip that can be programmed using a standard programming device and can be erased by the proper signal being applied to the erase pin. EEPROM is used primarily as a nonvolatile backup for the user program RAM. If the user program in RAM is lost or erased, a copy of the program stored on an EEPROM chip can be downloaded into RAM. It is common on some PLCs for the processor to load the program from the E^2PROM chip into RAM memory each time the processor is powered up or after a power failure. Figure 3–7 shows an EEPROM memory card used with the Modicon 984-120 Compact PLC to store the user program. This credit-card size device offers a convenient method for copying and/or loading user programs.

Figure 3–7 EEPROM Memory Card
(Courtesy of Modicon Inc.)

Memory Size

PLCs are available with memory sizes ranging from as little as 256 words for small systems up to 2 Meg (million) for the larger systems. Memory size is usually expressed in K values: 2K, 4K, 16K, and so on. K, or Kilo, which usually stands for 1,000, actually represents 1,024 in computerese. The difference between a standard K (1,000) and the 1,024K value used with processors and computers, is due to the way the words were counted. One of the counting, or numbering systems used with PLCs is the **binary system**. The binary system has a base 2, as contrasted to the decimal system we use every day that is base 10. Base 10 represents the numbers 0 through 9, which is 10 digits. The binary numbering system with a base 2 only has 2 digits. The digits are 1 and 0. As with the decimal system that has place values (tens, hundreds, thousands), the binary system also has place values. These are 1, 2, 4, 8, 16, and so on, and each place value is equal to twice the value of the previous number. Base 2^0 represents the number 1, base 2^1 represents 2, base 2^2 represents the number 4 ($2 \times 2 = 4$), base 2^3 represents the number 8 ($2 \times 2 \times 2 = 8$), and so on. Counting in this fashion, 2^{10} would equal 1,024. While 1,024 is actually larger than the 1,000 that K actually represents, K (with a value of 1,024) is used in PLCs, and personal computers as well. This also explains the reason for the odd memory sizes of individual memory chips: 256 (1 $\times 2^8$); and 512 (1 $\times 2^9$). A memory chip of 256 words would be $^1/_4$K and 512 words would be $^1/_2$K. A PLC with a total memory of 64K would actually have 65,536 words of memory (64 times 1,024). Words, word structure, and numbering systems are covered in Chapter 6.

While it is common for PLCs to measure their memory capacity in words, it is important to know the number of bits in each word. A PLC that uses 8-bit words would have half the memory capacity of a PLC that uses 16-bit words. For example, the PLC that uses 8-bit words has 65,536 bits of storage with an 8K word capacity ($8 \times 8 \times 1024 = 65,536$), whereas a PLC using 16-bit words has 131,072 bits of storage with the same 8K memory ($16 \times 8 \times 1024 = 131,072$). It is important to know the word size of any given PLC before memory size can be accurately compared.

NOTE: Personal computers and some PLC manufacturers, such as Simatic T.I., size memory in **bytes**, not words. A byte is 8 bits, or half of a 16-bit word. A 32-bit word would have 4 bytes.

The actual size of the memory required depends on the application. In the event that future expansion is planned, there are two options: buy a PLC with more memory than is presently necessary to allow for future expansion; or buy a PLC that meets present needs and add memory (upgrade) when the need arises. Depending on the manufacturer, adding memory may be as simple as replacement of the memory module, or it may require that additional memory chips be added to the existing memory module. Some processors have no provisions for memory expansion and must be replaced if the memory needs to be increased.

For PLCs in which the memory can be expanded by adding volatile RAM chip(s) to the memory module, the following procedures should be used
1. Record a copy of the current user program on disk or magnetic tape, depending on the system.
2. Remove main power from the PLC.
3. Remove the memory module and take to a clean area.
4. Carefully remove any screws necessary to gain access to the printed circuit board where the extra RAM sockets are located. If the back-up battery is located on the module, disconnect the battery before removing or installing memory chip(s).

NOTE: The RAM chip(s) will come packaged in a conductive plastic bag (often referred to as a "static" bag). Within the bag, each RAM chip will be inserted into a conductive sponge-like material. The conductive, yet highly resistive, material is used to keep all the pins of the chip at the same electrical potential.

    * CAUTION: When working with the RAM chips, *do not* handle cellophane covered articles such as cigarette packages or candy wrappers, plastic, styrofoam, or other materials that can cause a static charge. *Do not* install the chip in carpeted or contaminated areas where pins may become fouled. And *never* slide the RAM chip across any surface, store a RAM chip in a nonconductive plastic bag, or insert the chip into nonconductive material.

The volatile RAM chips used today are not as susceptible to damage from static charges as they were just a few years ago. But rather than just removing the chip from the conductive material and installing it into the proper socket, the following precautions should still be used
1. Ground all tools before contacting the RAM chip.
2. Wear a conductive wrist strap which has a minimum 200K ohm resistance and is connected to earth ground as shown in Figure 3–8.
3. Control relative humidity at 40%–60%, if possible.

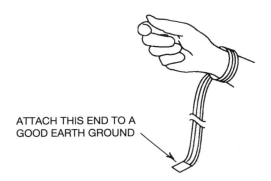

ATTACH THIS END TO A
GOOD EARTH GROUND

Figure 3–8 Wrist Strap Grounding Device
(Courtesy of GE-Fanuc Automation)

Remove the chip from the conductive foam. Be careful to touch only the chip base. *Do not* touch the pins. Inspect the pins for proper alignment. If any pins have been bent, gently straighten them using needle-nose pliers that have been grounded. A dot or notch on the case of the chip is used for proper orientation of the chip into the socket. Grasp the chip by both ends and gently set it in the socket. *Do not insert.* Be sure the chip is positioned so the dot or notch of the chip matches the dot or notch on the socket.

Before attempting to insert the chip into the socket, check each pin to make sure it lines up properly with the corresponding socket point. Make any necessary pin adjustments as outlined above.

When pin alignment is ensured, insert the chip into the socket. Insertion is accomplished by pressing *gently* on the case of the chip until the chip is fully seated into the socket.

Carefully reassemble the memory module, remembering to reconnect the back-up battery if one was mounted on the module. Reinstall the module in the processor and reapply power to the system. The user program can now be reentered into the processor and any additional user program can be added using the new memory chip(s).

Memory Structure

As indicated earlier, the processor memory is divided into two general classifications: **user memory** and **storage memory**.

User memory contains the ladder diagram instructions programmed by the user. The instructions are entered either by a programming device, hand-held or desktop-type, a magnetic tape, or a system computer. Programming can also be accomplished by using a telephone interface from a remote computer, tape loader, or programming device. The user memory may also store user programmed messages that are recalled or activated by contacts in the ladder diagram. Upon activation, the message(s) is displayed on the VDT of the programming device or other remote VDT, or can be printed out on a compatible printer.

Message storage and recall can be used to alert an operator if a bearing is heating up. A typical programmed message might be: "Bearing hot on Motor #4." A thermocouple at the bearing of Motor #4 would be tied to one input circuit of a thermocouple input module. When the temperature of the bearing exceeds a predetermined value, the thermocouple causes that circuit of the thermocouple input module to close, or turn *ON*. The corresponding contact of the input device programmed in the ladder diagram closes, and the message is generated.

Storage memory is where the status (*ON* or *OFF*) of all input and output devices are stored. Numeric values of timers and counters (preset and accumulated), numeric values for arithmetic instructions, and the status of internal relays are also stored in this memory.

While the information presented in this section applies generally to all PLCs, more specific information and memory structure can only be obtained by reviewing the specifications and literature of individual manufacturers. In subsequent chapters, the memory structure of specific PLCs will be discussed and illustrated, but the text does not cover all the PLCs on the market today.

Peripherals

Peripherals are defined as devices connected to a processor that provide an auxiliary or support function.

A printer (Figure 3–9) for producing hard copy printouts of the user program and/or other processor information is an example of a peripheral. The printer may be activated by a keyed sequence entered from the programming device and/or initiated from the processor itself. The mode of printer communication (serial or parallel) and the cables that are required vary, based on the hardware and/or software used. The connector for connecting the printer may be mounted on the back of the programming device, on the processor, or both. Figure 3–10 shows a connector on the back of a programmer. To avoid the frustration of having the printer and processor not communicate with each other, consult the installation manual(s) for the PLC that is being installed.

Figure 3–9 Dot Matrix Printer
(Courtesy of OKIDATA)

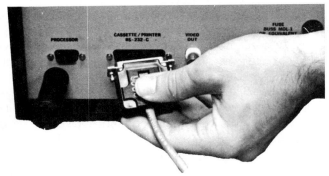

Figure 3–10 RS-232C Connection for Printer/Cassette
(Courtesy of Square D Company)

Using a compatible computer terminal and the appropriate software enables the user program and data files to be stored on floppy disk and/or the hard drive. Figure 3–11 shows an Allen-Bradley T60 industrial terminal.

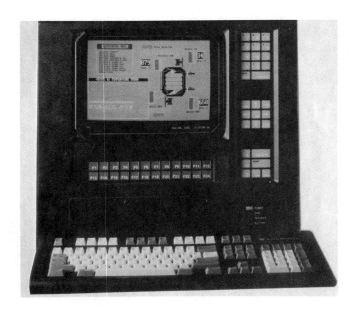

Figure 3–11 Allen-Bradley LT-60 Industrial Terminal
with $3^1/_2$" Floppy Disk and Hard Drive
(Courtesy of Allen-Bradley)

Another peripheral device is the magnetic tape loader. The tape loader is used to record and store the user program or to load preprogrammed instructions into the processor. Data quality cassette tapes or data cartridges are used by the tape loader, depending on the type, to record and store the user program. Recording the user program provides a back-up program in the event the processor program is lost due to memory failure or accidental erasure.

A standard cassette recorder can be used with some smaller PLCs for recording and loading programs. Expense is greatly reduced when an ordinary cassette recorder or player can be used.

Dedicated tape loader and cassette recorders, while still in use in many plants, are not as popular as they were just a few years ago. The increased use of E^2PROM, UVPROM, and EAROM memory chips, as well as the increased use of the personal computer for programming PLCs has eliminated the need for tape loaders, because the program can automatically be stored on the memory chip, or stored on the computer floppy disk or hard drive.

Another peripheral device is the **MODEM** (**MO**dulator and **DEM**odulator). The MODEM can be used to connect the processor via telephone lines to a programming device, computer, tape loader, printer, or to another PLC. Figure 3–12 shows a typical MODEM hookup from a PLC processor to a remote programming device.

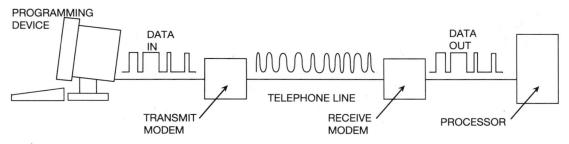

Figure 3–12 Typical MODEM Hookup

Chapter Summary

The processor contains the circuitry necessary to monitor the status (*ON* or *OFF*) of all inputs and control the condition (*ON* or *OFF*) of all outputs. It also has the ability to solve and execute the individual program steps in the user program, has a memory for storing the user program and other numeric information, and has the ability to retrieve and use any and all information stored in memory. The processor not only communicates with I/O racks and the programming device, but also communicates with any peripheral device(s) that may be connected to it. The memory used in PLCs is of two distinct groups: volatile and nonvolatile. Volatile memory requires a battery back-up to prevent the program from being lost due to a power failure; nonvolatile memory holds the program when power is lost or turned off. Programs can be stored on various types of memory chips, as well as on floppy disks or on the hard drive of a computer programmer.

Peripherals such a printers, tape loaders, and MODEMs are devices that can be connected to the PLC to provide various support functions.

REVIEW QUESTIONS

1. The processor is often referred to as the _____ of the programmable controller.
2. Briefly describe *volatile memory*.
3. Briefly describe *nonvolatile memory*.
4. 1K of memory is actually
   a. 1,000 words
   b. 1,010 words
   c. 1,024 words
   d. 1,042 words
5. Calculate the actual number of words in an 8K memory.
6. The most common type of volatile memory is
   a. PROM
   b. EAROM
   c. UVPROM
   d. RAM
7. Which of the following are types of nonvolatile memory?
   a. EEPROM
   b. PROM
   c. RAM
   d. EAROM
   e. UVPROM
8. List the two broad categories of memory (not volatile and nonvolatile).
9. Define the word *peripheral*.
10. List two common peripheral devices.
11. What is a *watchdog timer*?
12. What special precautions should be taken with lithium batteries?

Chapter 4

Programming Devices (Programmers)

Objectives

After completing this chapter, you should have the knowledge to
- Describe the function of a programming device.
- Describe the three types of programming devices.
- List advantages and disadvantages of the three types of programmers.
- Explain the term *on-line programming*.

Programming Devices

A programming device is needed to enter, modify, and troubleshoot the PLC program, or to check the condition of the processor. Once the program has been entered and the PLC is running, the programming device may be disconnected. It is not necessary for the programming device to be connected for the PLC to operate, but it can be used to monitor the PLC program while the program is running.

Programming devices, or programmers as they are most often called, come in three types: dedicated desktop, hand-held, and computer (Figures 4–1a, 4–1b, and 4–1c).

Figure 4–1a Hand-Held Programmer
(Courtesy of Modicon Inc.)

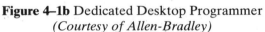

Figure 4–1c IBM® Laptop Computer

Figure 4–1b Dedicated Desktop Programmer
(Courtesy of Allen-Bradley)

Each type of programmer has advantages and disadvantages. After all three types have been described, the merits of each will be discussed.

Dedicated Desktop Programmers

Dedicated desktop programmers like the one pictured in Figure 4–1a are primarily designed for programming and monitoring the PLC. They are not capable of performing other computer functions like running software programs for word processing and spreadsheets, but are, however, designed to be portable and withstand the mechanical shock associated with moving the programmer from job site to job site. They are also designed to function in the industrial environment where there is electromagnetic noise, high temperatures, and humidity.

A typical dedicated programmer consists of a keyboard, VDT (video display terminal), and the necessary electronic circuitry and operating memory for developing, modifying, and loading a program into the processor memory.

Keyboard   The keyboard of the dedicated programmer may have raised keys, like a standard computer terminal (Figure 4–2), or have a flush, sealed touchpad-type keyboard (Figure 4–3). The flush, sealed touchpad keyboard is ideally suited for environments that are dusty or have metallic particulates in the air.

Figure 4–2 Programmer with Raised Keyboard
(Courtesy of Square D Company)

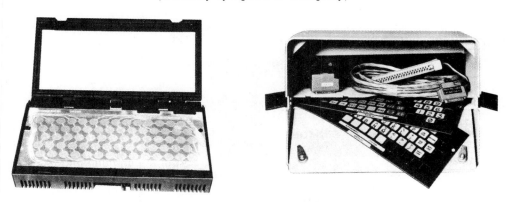

Figure 4–3 Sealed Touchpad Keyboard and Keyboard Overlays
(Courtesy of Allen-Bradley)

Keyboards on dedicated programmers vary quite a bit, depending on the manufacturer. Most dedicated programmer keyboards have electrical symbol keys for normally open contacts, normally closed contacts, timer contacts, and the like. The various symbols are also referred to as **instructions**. The normally open contact symbol key, when pressed, is an instruction to the processor as are all the other symbol keys (i.e., the output key is an output instruction; the timer contact is a timer instruction). These dedicated keys make programming easier for those electricians or technicians who are intimidated by computers or computer-like devices. The keyboard also has special function keys that are used for program development, along with numeric (number) keys for addressing the various components used in the circuit. The address of a device tells the processor the type of device it is (input or output) and its location. Most keyboards also have alpha (letter) keys for writing program notations, labeling the devices in the program, report generation, and other special programming functions.

Video Display Terminals   Video display terminals give the user a visual display of the program and range in screen or tube sizes from 5 inches to 12 inches, measured diagonally, with the video display usually being black and white or green on green. The video screen allows the user to display multiple lines or Rungs of the program. The ability to view several lines of the program at once is a powerful troubleshooting aid.

Video brightness and contrast adjustments are found either on the front or rear of the programmer. A video jack is available on the rear of most programmers for connecting an additional video monitor. VTDs are sometimes referred to as CRTs (cathode ray tubes).

Once the program has been entered into user memory, the PLC is ready to control the circuit. When the processor is placed in the *RUN* mode and the circuit activated, the programmer VDT gives a visual display of the circuit condition.

Actual circuit condition is shown on the VDT display in basically two ways: some PLCs intensify, or make brighter, all contacts, interconnecting lines, and coils that are passing current or have power flow; others intensify or use reverse video only to indicate which contacts and coils have power flow. Figure 4–4a illustrates how a circuit appears before the *START* button is pushed for a system that intensifies contacts, interconnecting lines, and coils; Figure 4–4b shows the VDT display after the *START* button is pushed and the holding contacts close.

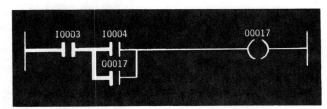

Figure 4–4a VDT Display Prior to Start Button Being Depressed

Figure 4–4b VDT Display After Start Button is Depressed and Holding Contacts Close

The terms "Passing current" or "power flow" are hold-overs from hard-wired circuits. In fact, there is no power flow, or current flow, as we normally think of it; rather, it is logic continuity or logically true statements. The logic of the circuit shown in Figure 4–4a is as follows; if input I0003 and input I0004 are closed or *true*, then output O0017 should be turned *ON* or go true. When output O0017 goes true, the instruction labeled O0017 (holding contacts) will become true, and an alternate logic statement is now true. When I0003 and O0017 are true, output O0017 will remain true or *ON*, even if input I0004 opens, or goes *false*.

Figure 4–5a illustrates how a display using reverse video looks before the *START* button is pushed, and Figure 4–5b shows the display after the *START* button is depressed and the holding contacts close.

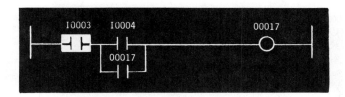

Figure 4–5a Reverse Video Display Prior to Start Button Being Depressed

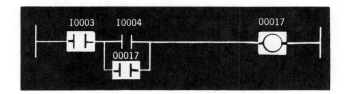

Figure 4–5b Reverse Video Display After Start Button is Depressed and
Holding Contacts Close

No matter which method is used, this feature of dedicated desktop programmers is a powerful troubleshooting aid. By viewing the display on the VDT, the electrician or technician can determine which contacts are closed and which outputs are turned on.

NOTE: An output coil that is intensified only indicates that the output module circuit is *ON*. It does not guarantee that the actual output device is *ON*. However, if the output device is all right, it will be *ON* any time the output module circuit is *ON*.

To provide a dependable back-up of the program in case the memory fails or is inadvertently cleared or altered, an E^2PROM chip or a cassette tape is used to record and/or load the user program. Data quality tapes should be used to ensure accurate transfer of data.

Hand-held Programmers
Hand-held programmers are smaller, cheaper, and more portable than dedicated desktop programmers. While the portability is a real plus, the hand-held programmer has some limitations.

Unlike the desktop programmer that can display a complete circuit network on the VDT, hand-held programmers have limited display capabilities. Some hand-held programmers display a Rung of logic with up to four horizontal lines, while others only display one line or one element at a time. The display is either LEDs or liquid crystals. Figure 4–6 shows an Allen-Bradley hand-held programmer with liquid crystal display.

Figure 4–6 Hand-Held Programmer
(Courtesy of Allen-Bradley)

The hand-held programmer usually does not have the full programming features of the dedicated desktop programmer, and requires more "keystrokes" to actually enter a program. Hand-helds typically have restricted access to the processor memory.

On the plus side, hand-held programmers are well-suited for installations that require constant changes in circuit requirements because they are light-weight (normally less than 2 pounds), portable, and ruggedly constructed. It is much easier to connect the hand-held programmer to the processor for changing program parameters or for troubleshooting than it is to bring out the large, heavier desktop programmer.

While the relatively low cost of hand-helds make them affordable troubleshooting tools, it takes more time to go through the program one contact or Rung at a time. The extra time is the trade-off for the lower initial cost of the programming device.

Computer Programmers

With software available for most major brands of PLCs, the trend is to use a personal computer for programming. The personal computer is lower in cost than a dedicated desktop programmer, and has the added feature of being able to perform functions other than programming a PLC.

Software has been developed by the PLC manufacturers themselves and also by other companies. Figure 4–7 shows a personal computer being used to run Modicon Inc. Modsoft® software.

Figure 4–7 Personal Computer Running PLC software
(Courtesy of Modicon Inc.)

When software for a specific brand of PLC is developed by a company other than the manufacturer of that PLC, the software is referred to as "third party software." These software programs can generally be used with most IBM® compatible computers in a DOS® environment, or with other popular software programs, like Windows®, to increase their effectiveness and flexibility.

For a personal computer to run the various software packages that are used for programming PLCs, the following are the minimum general hardware requirements that must be met:
- IBM® PC/XT, AT or compatible.
- 640K RAM.
- DOS 3.0 or greater.
- Hard disk with 4 megabytes (MB) of available disk space.

Using a personal computer for programming provides many of the advantages of the dedicated desktop programmer, but also provides features not available on most dedicated desktop programmers.

The personal computer usually has a color monitor (VDT), and the monitor shows multiple Rungs of program logic, as well as highlighting the circuit elements to indicate status, just like the dedicated desktop programmer. The computer has the added ability to interface the PLC software program with other software programs for "cut and paste" program development and editing. PLC software usually provides more documentation capabilities than does the dedicated desktop processor. The documentation may be in the form of labeling each element, or writing

Rung comments. Added graphic capabilities are also normally a part of a PLC software program. The addition of graphics allows the electrician or technician to develop flow diagrams of the controlled equipment (Figure 4–8), provide operator alarms and messages, and utilize other useful process information.

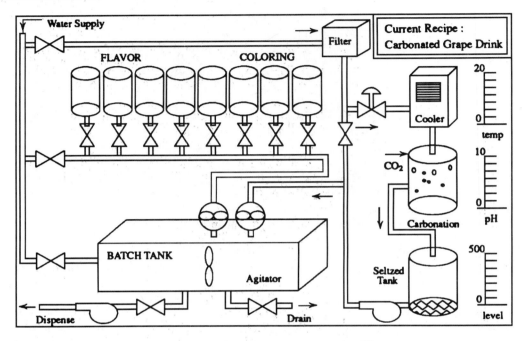

Figure 4–8 Flow Diagram Using ControlView™ Software
(Courtesy of Allen-Bradley)

A personal computer is relatively low in cost when compared to a dedicated desktop programmer. Some PLC manufacturers and software suppliers, however, require a communication card to run their software, and this requirement adds to the initial cost. The communication card must be installed in an empty slot in the personal computer, and without the card, the software will not communicate with the processor. If the software was copied and loaded into another computer, the computer could not communicate with the PLC without an additional communication card. Although the software could be used to develop a program on a computer without a communication card, the program would have to be transferred to the computer with the card to run the program. When a program is first developed and edited, it is done in the program mode or the **off line** mode. Off line indicates that the program has not yet been loaded into the processor. The program is not operational until it is loaded into processor memory, and the processor is placed in the *RUN* mode. Once the program has been checked, the program is loaded into the processor for testing and further verification and/or modification. Changes that are made after the program has been loaded into the processor and the I/O section has been activated, is called **on line** programming.

*CAUTION: Making changes to the program while the program is running and the driven equipment is operational (on line programming) must only be done by trained personnel who not only understand the PLC program, but also thoroughly understand the driven equipment and/or process.

Other manufacturers use what is called a **hardware key** to prevent unauthorized use of their software programs. The hardware key is usually connected to one of the printer ports. When the software is first turned *ON*, the software looks for the hardware key. If the key is not installed, the software will not function. Still other manufacturers do not copy-protect their software, and, in fact, encourage copying and widespread use of the software.

Personal computers are also available in the small laptop variety (shown in Figure 4–1c) or the even smaller notebook style. These small computers add to the flexibility of using a computer as a programmer.

Using Personal Computers for Programming—Advantages   A major advantage of using a personal computer for programming is the ability of the personal computer to store the program on floppy disk and/or on the hard disk. If for some reason the program is lost, the restoration of the program is simple. Merely copy the program from disk to the processor memory.

When the software is updated by the manufacturer to provide additional features, the update is easily accomplished by loading the new software program onto the computer hard disk by way of the floppy disk drive. A dedicated desktop programmer requires changes in ROM chips to accomplish updates.

When troubleshooting, the software that is used for programming a PLC from a personal computer has the added feature of being able to display Rungs of logic in any order that the electrician or technician may find helpful. On a large program, for example, Rungs 4, 45, 46, 67, and 75 are displayed on the screen at one time. This feature, while available on a few dedicated desktop programmers, is standard on most software programs that have been developed for the various PLCs on the market today.

While dedicated desktop programmers can be networked with multiple PLC processors, they do not have the flexibility and networking capabilities of a personal computer.

Using Personal Computers for Programming—Disadvantages   One disadvantage of using a personal computer for programming is that the programming is not as user- or electrician-friendly as a dedicated desktop programmer with its symbol keys. It's not that the programming is hard using a computer; it just takes longer to learn how the programming is done. In other words, the learning curve is longer. Another, and probably the biggest disadvantage of using a personal computer as a programmer, is that the computer is not designed for the industrial environment. The personal computer is affected by electrical noise, and high and low temperatures as well as high humidity, conditions that dedicated desktop programmers are designed to handle.

The ideal programmer, then, appears to be a mix of the dedicated desktop programmer and the personal computer. That is exactly what a number of the PLC manufacturers thought, so they developed industrial-rated computer programmers. Allen-Bradley, Siemens T.I., Modicon Inc., and GE-Fanuc Automation are several of the companies that offer these highbred programmers. Figure 4–9 shows the Modicon Inc. P230 programming panel.

Figure 4–9 Modicon Inc. P230 Industrial Programming Panel
(Courtesy of Modicon Inc.)

The P230 is a 25 Megahertz (Mhz) 486SX computer with a 120MB self-parking hard drive. Self-parking means that the computer automatically parks the hard drive each time it is shut off. Parking the hard drive prevents damage during transportation of the computer. The P230 has a 4Meg RAM that is expandable to 32Meg and comes with two serial communication ports, one parallel port, and a VGA monitor port.

As might be expected, this type of programmer is more expensive than the dedicated desktop programmer or a personal computer. For the money, though, one gets the best features of both. Is the added cost justifiable? That question can only be answered after all the variables have been considered. The variables could include, but not be limited to:
- Environment.
- Size of program.
- Graphic requirements.
- Requirements for other software, i.e., spreadsheets, word processing.
- Programmability.
- Networking ability.

A final review of some of the advantages and disadvantages of each type of programmer follows.

Dedicated Desktop Programmers

Advantages
- VDT to display multiple Rungs of the program.
- Ease of programming, user-friendly.
- Designed for industrial use.
- Portability.
- Status of circuit highlighted (intensified or reverse video).

Disadvantages
- Relative high cost.
- Limited models of PLC that can be programmed.
- Limited documentation and graphics capabilities.
- Physical size.

Hand-held Programmers

Advantages
- Low cost.
- Small size.
- Very portable.

Disadvantages
- Can only view limited Rungs or instructions.
- Fewer functions.
- Limited access to memory.
- Limited or no documentation capabilities.
- Works with only a specific PLC model.
- More difficult to program than a dedicated desktop programmer.
- Uses logic status lights instead of intensified or reverse video.

Personal Computer Programmers

Advantages
- VDT to display multiple Rungs of logic.
- Can program any type of PLC (with appropriate software).
- Extensive documentation capabilities; Rung comments, labels, and so forth.
- Cut and paste editing capabilities.
- Low cost when compared to a dedicated desktop programmer.
- Small and portable when laptop or notebook type are used.
- Easy to make copies of program on floppy disk or hard drive.
- Software can be upgraded without changes in hardware.
- View multiple Rungs in any numeric order for troubleshooting.

Disadvantages

- Must operate in normal environment, not industrial rated.
- Initial software costs.
- Software updates or license renewal costs.
- More difficult to program, longer learning curve.
- May require special communication card.

In the final analysis, if money is not a consideration, or the specific application warrants the difference in cost, the industrial computer programmer is the most versatile and serviceable of all the types of programmers that are available.

Chapter Summary

The programming device, or programmer, is used to enter, modify, and monitor the user program. Which type of programming device to use will vary with each application, and what is appropriate for one application may not be the most appropriate for another. The program (ladder diagram) is entered by pushing keys on the keyboard in a subscribed sequence, with the resultant circuit(s) displayed on the VDT of a desktop programmer, and with a LED or liquid crystal display for the hand-held programmer. The visual display can be used as an aid to test the circuit prior to entry into user memory, or for troubleshooting after the circuit is entered into memory and is operational. Contacts and coils are either intensified or displayed in reverse video to indicate power flow or logic continuity. Programming the PLC is not difficult, but time must be spent to become familiar with the specific PLC and its programming techniques.

REVIEW QUESTIONS

1. Briefly describe the function of the *programming device*.
2. What does the acronym VDT stand for?
3. Define the term *on line*.
4. List the three type of programming devices and give advantages and disadvantages for each.
5. Define the term *off line*.

Chapter 5 Memory Organization

Objectives

After completing this chapter, you should have the knowledge to

- Identify the two broad categories of memory and describe the function of each.
- Identify the types of information stored in each category of memory.
- Define the term *byte*.
- Define the acronym *bits*.
- Define *holding registers*.
- Understand the term *default*.

Memory Words and Word Locations

For the programmable controller to function properly and control a process or driven equipment, it must be able to perform the user program repeatedly and accurately. The system must also be able to perform its control function with great speed, which is achieved by processing all information in binary signals. The key to the speed with which binary information can be processed is that there are only two states, each of which is distinctly different. The binary signal falls into one of two states which are 1 and 0. The 1 and 0 can represent *ON* or *OFF*, true or false, voltage or no voltage, high or low, or any other two conditions depending on the system. There is no in-between state or condition, and when information is processed, the decision is either *yes* or *no*. There is no *maybe*, *almost*, or any other alternative.

As indicated in Chapter 3, the processor memory consists of hundreds or thousands of locations. that are referred to as words. Each word is capable of storing binary data in the form of binary digits, or **bits** (**BI**nary digi**TS**). A binary digit, like a binary signal, can only be a 1 or a 0. The number of bits that a word can store will depend on the system or PLC. Words can be made up of 32 bits, 16 bits, or 8 bits. The 16-bit word is the most common (Figure 5–1).

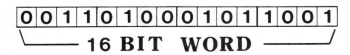

Figure 5–1 16-Bit Word

If a memory size is 256 words, then it can actually store 4,096 bits of information using 16-bit words (256 words × 16 bits per word) or 2,048 bits using an 8-bit word (256 words × 8 bits per word). When comparing memory sizes of different PLC systems, it is important to know the number of bits per word of memory. Bits can also be grouped within a word into **bytes**. A byte is a group of 8 bits.

So that information stored in each word can be located, each word is numbered or given an **address**. Addressing words in the memory serves the same function as the addresses used for homes or apartments. Word 100, for example, represents a specific word location in memory, just like N. 100 Lincoln represents the address of an apartment building. The bits in word 100 are found by referencing a given bit number, just like the occupant of the apartment complex is found by a given apartment number.

Since a bit of information can only be a 1 or a 0 (*ON* or *OFF*), how is the status of bits within a word determined? Words that store the status of individual bits for input devices are set to 1 (*ON*) or 0 (*OFF*), depending on the status (*ON* or *OFF*) of the input devices that the bit locations represent. Other bits are set to 1 or cleared to 0 by the processor in response to the logic of the user program, RELAY LADDER LOGIC, or special instructions, which, in turn, control the status (*ON* or *OFF*) of other bits that represent output devices.

A simple example of how this works is illustrated in Figure 5–2.

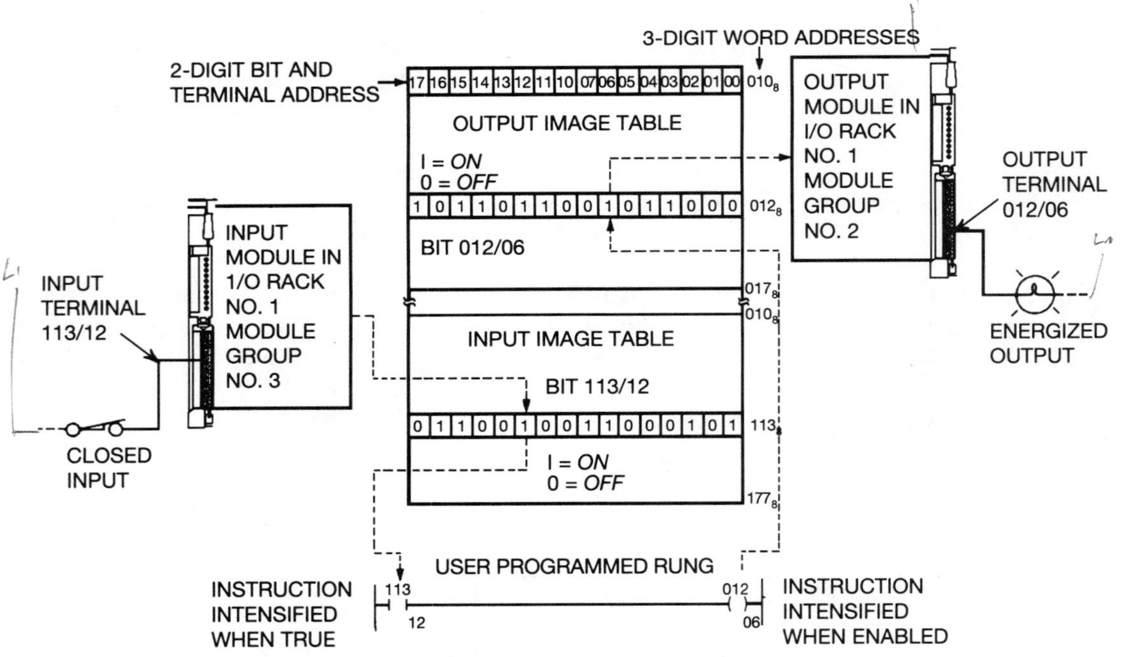

Figure 5–2 Relationship of Bit Address to Input and Output Devices
(Courtesy of Allen-Bradley)

NOTE: The example uses memory organization and addressing utilized by the Allen-Bradley PLC-2 and PLC-5 families. While the example is specific to Allen-Bradley, the concepts illustrated are common to all PLCs. Allen-Bradley uses an octal numbering system to address bit locations. Notice that the 16 bits are numbered 00 through 07 and 10 through 17. In the octal numbering system, the numbers 8 and 9 are never used. The octal numbering system will be covered in detail in Chapter 6.

Assume that when a given limit switch is closed, the closure will turn an indicator lamp *ON*. The limit switch is connected to an input module in the I/O rack, while the indicator lamp is connected to an output module. Chapter 2 discussed DIP switches that were set in a prescribed sequence to identify the I/O rack number for the processor, and that the location of each terminal point of each I/O module within the rack determined the address of a given device. In Figure 5–2, the limit switch is connected to terminal 11312 on an input module or has an address of 11312. This indicates that bit 12 of word 113 stores the status (*ON* [1] or *OFF* [0]) of the limit switch. The indicator lamp is connected to terminal 01206. This address indicates that bit 06 of word 012 controls the status (*ON* [1] or *OFF* [0]) of the lamp.

By programming a simple circuit into the user memory of the processor (shown at the bottom of Figure 5–2), the processor controls the indicator lamp using the logic of the user program. The logic states that if contact 11312 closes, lamp 01206 should light, or go *ON*. When power is applied to the processor, the processor starts its scan and looks at bit 12 of word 113 to see if the bit is set to 1 or 0. If the limit switch is open, the bit will be set to 0, or *OFF*. If the limit switch is closed, as indicated in Figure 5–2, the input module sends a signal to the processor, and bit 12 of word 113 will be set to 1, or *ON*.

The next part of the scan solves the user program. The logic of the ladder diagram indicates that when contact 11312 (bit 12 of word 113) is closed, or *ON*, the indicator lamp 01206 should be turned *ON*. The processor reads the logic, and during the third step of the scan, sets bit 06 of word 012 to 1, which turns the lamp connected to the output module *ON*.

The address 11312 also tells us that the limit switch is an input device, and is wired to terminal 12 of module group 3 of rack 1.

Allen-Bradley uses 5-digit addresses with its PLC-2 family of programmable logic controllers. Figure 5–3 illustrates the significance of each digit or group of digits.

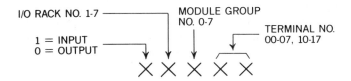

Figure 5–3 Allen-Bradley 5-Digit Address Format

The first digit is used to indicate whether the address is an input or output. The number 1 represents an input device, while a 0 represents an output device.

The next, or second digit, identifies the rack number. This is always a number from 1 through 7 (octal numbering). The next number identifies the module group within the rack. This is always a number from 0 through 7. The last two digits identify the actual terminal number that the device is wired to.

Figure 5–4 reviews the concept using the address of the limit switch 11312.

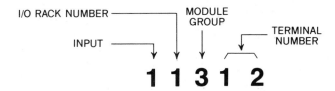

Figure 5–4 Limit Switch Address 11312

A 1 in the first digit location tells us that the address represents an input device. The next digit, also a 1, tells us that the device is located in I/O rack number 1. The next digit, which is a 3, further identifies the location as module group number 3. The last two digits, 1 and 2, identify the actual terminal (12) on the input module that the limit switch is connected to.

Another example of this concept is shown in Figure 5–5. The limit switch address 11312 gives us a hardware location for an input device in rack 1, module group 3, terminal 12. This same address, 11312, tells us that the status (*ON* or *OFF*) or state of the limit switch is reflected by bit 12 of word 113.

This same addressing scheme gives us a hardware location for the indicator lamp addressed 01206. The 0 in the first digit location indicates an output device. The next digit, a 1, tells us that the I/O rack location is rack 1. The next digit identifies the module group as group 2. The last digits locate terminal 06 as the terminal on the output module that the indicator lamp is wired to. Again, the address 01206 also locates the memory word and bit location that reflects the status (*ON* or *OFF*) of the indicator lamp as shown in Figure 5–6.

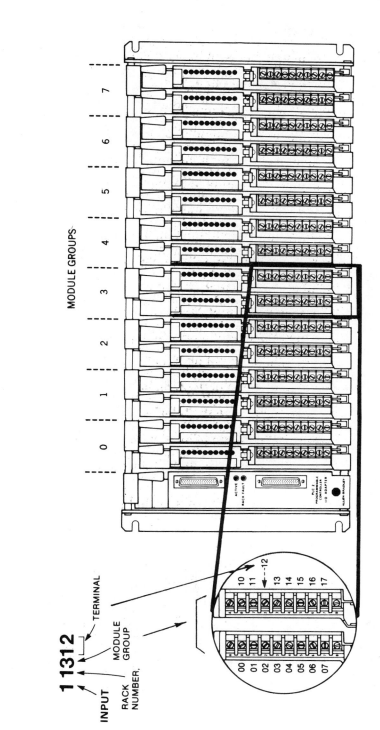

Figure 5–5 Relating Input Address 11312 to Actual Hardware Location
(Courtesy of Allen-Bradley)

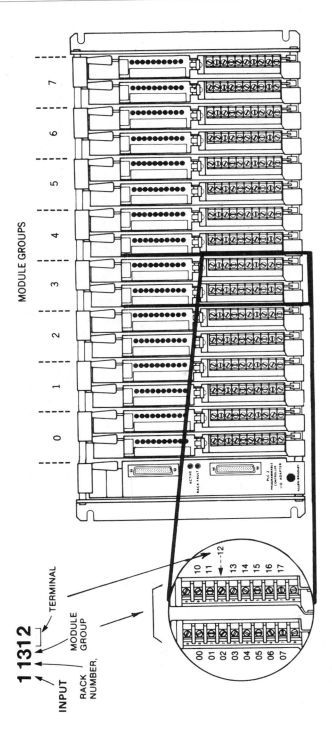

Figure 5–6 Relating Output Address 11312 to Actual Hardware Location
(Courtesy of Allen-Bradley)

For larger systems that use more than seven I/O racks, a 7-digit number is used as shown in Figure 5–7.

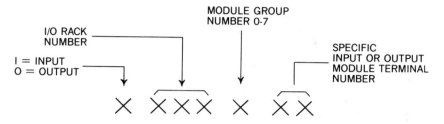

Figure 5–7 Allen-Bradley 7-Digit Address Format

For their PLC-5 family of controllers, Allen-Bradley uses the addressing scheme shown in Figure 5–8.

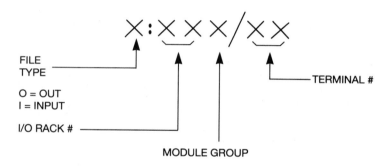

Figure 5–8 PLC-5 Addressing Structure

Figure 5–8a shows an address that indicates it is an output address, and that the output device is located in rack 01, module group 2, and is connected to output terminal 01.

0:012 / 01

OUTPUT, RACK 1, MODULE GROUP 2, TERMINAL 01

FIGURE 5–8a Example of Output Address

While the address system discussed is specific to Allen-Bradley, most PLC manufacturers use an addressing scheme that identifies memory word locations, and may also give hardware locations.

Memory Organization

At this point, it should be no surprise to find out that not all PLC manufacturers have organized their memories in the same way, or that they do not all use the same terminology for the configuration or make-up of their memories.

As discussed in Chapter 3, there are two general classifications of memory: **storage memory** and **user memory** (Figure 5–9).

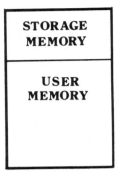

Figure 5–9 Two Broad Categories of Memory

Storage Memory

Storage memory is that portion of memory that will store information on the status of input and output devices, preset and accumulated values of timers and counters, internal relay equivalents, numerical values for arithmetic functions, and so on. The entire storage memory is called a **data table**, a **register table**, or other names, depending on the PLC manufacturer. A register is defined as an area for storing information (logic or numeric). Although the names or titles which are given to sections or subsections of the storage memory vary, the principles involved do not.

For example, the section of the memory that stores the status of the real-world input devices may be referred to as an input image table, input register, input status table, or external input section. No matter what name is used, the information is stored in the same way. The status (*ON* or *OFF*) of each input device is stored as either a 1 or a 0 (*ON* or *OFF*) in one bit of a memory word. When the processor is executing the user program (ladder diagram), it scans the input device status stored in the storage memory to determine which inputs are *ON* or *OFF*.

The section of storage memory set aside for output status may be referred to as the output image table, output register, output status table or external output section. Again, the name does not change the function of this section of the storage memory, or the method by which information is placed in memory for control of the actual output devices. As the processor executes the user program, it sends binary data (1s or 0s) to the output section of memory to control the output devices. Each output device is represented by one bit of a memory word.

Numeric information for timer or counter preset and accumulated values, arithmetic functions, sequencer functions, data manipulation, etc., uses a part of the storage memory that is called data registers or internal storage. Information is entered and stored in this part of memory using the binary, BCD, or hexadecimal numbering systems (the various numbering systems are covered in Chapter 6). The numbering system(s) used depend on the PLC hardware and system requirements. The storage of numeric information requires that several bits be used of one word to represent numbers. In a practical sense, any word used to store numerical information is not available for additional storage, even if all the bits of word are not used.

In general, any unused bits of a word that are not used for storing numeric values can be used as internal relays. Internal relays will replace the numerous control relays used in most hard-wired control circuits. Many PLCs have a portion of memory set aside just for internal relays. Internal, or dummy relays, may also be programmed in the output section of memory when all the words or bits of words are not being used for real-world output devices. The concept and use of internal, or dummy relays, is covered later in the text.

Figure 5–10 shows the address table for the storage memory section of a Square D Company Model 300 processor.

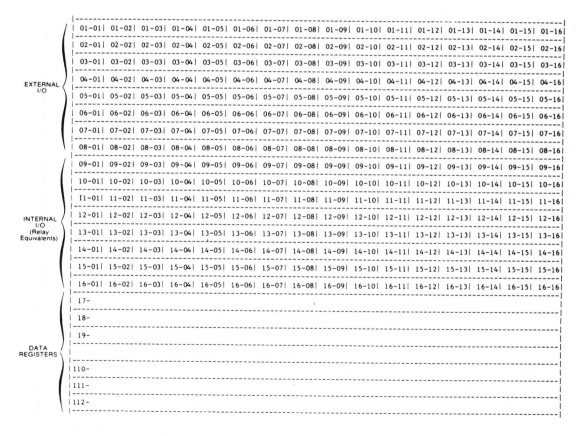

Figure 5–10 Square D Model 300 Storage Memory Addresses

The discrete inputs and outputs would be addressed 01-01 (word one-bit one) through 08-16 (word 8-bit 16) for a total of 128 I/O. Words 9 (bits 1–16) to word 16 (bits 1–16) are for internal I/O (relay equivalents), and words 16–112 are numerical data storage. While unused external I/O words can be used for storage, unused storage words cannot be used for external I/O.

The complete memory for the Model 300 also has user memory and additional memory for processor use.

Figure 5–11 shows the memory organization for an Allen-Bradley Mini PLC-2/15. The 2/15 is part of the PLC-2 family of Allen-Bradley processors. Allen-Bradley also has the newer PLC-5 family of processors. While the PLC-2 family is the older version, there are literally thousands of this type still in use in industry today, and an understanding of their memory structure is meaningful.

For the PLC-2 processors, Allen-Bradley uses 16-bit words. The word and bit addresses are numbered using the octal numbering system, as shown in Figure 5–11. The first eight words (000–007) of the data table are set aside for the processor. Words 010–017 and 020–026 are the output image table. An output device addressed 01000 is bit 00 of word 010; similarly, an address of 02617 is bit 17 of word 026. Note; word 027 is used by the processor for a *BAT LOW* condition, message generation, and data highway. The next 40-word section of memory, words 030–037, 040–047, 050–057, 060–067, and 070–077, is used for storing timer or counter accumulated values (numeric) or for internal storage.

The next 8 words are the second processor work area. Words 110–127 (16 words) make up the input image table for discrete input addresses. Input address 12406 is bit 06 of word 124. Words 125 and 126 are used by the processor to indicate remote I/O faults. Words 130–177 (40 words) are used to store the preset values of timers or counters or for internal storage. If word 030 is used for a timer, word 130 automatically stores the preset value of timer 030. The accumulated value for timer 030 is stored in word 030. Any timer or counter in an Allen-Bradley PLC-2 system will store the preset value in the word that is numbered 100 higher than the timer or counter number. For example, counter 047 has its preset value stored in word 147.

Below the timer or counter preset values and internal storage section is the user program section of memory.

The actual configuration of the data table can be changed to meet user needs. Input/output image tables can be increased or expanded to handle more discrete inputs and outputs. Additional information on expanding the data table of Allen-Bradley PLCs is available from their local representative or from their technical publications.

User Memory

The user memory, or logic memory as it is sometimes called, is where the programmed ladder diagram is entered and stored. Within the user memory, words are set aside as **holding registers**. Holding registers typically store information generated and used by the processor when it is solving the user program. Holding registers that are set aside to store intermediate values or other short-term bits of information are sometimes referred to as *scratch areas* or *scratch pads*.

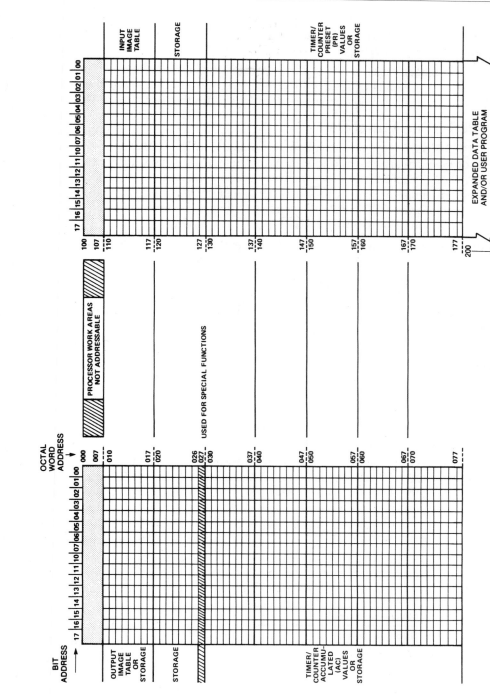

Figure 5–11 First 128 Words of an Allen-Bradley PLC-2/15 Memory

The user memory accounts for most of the total memory of a given PLC system. A system with an 8K memory (8192 words) typically has a storage memory of 2K or less, and the balance of memory (6K) is available for user memory.

Once the user program has been entered into the user memory, by either a programming device, tape loader, computer, or by means of a telephone interface, the programmable controller is ready to control the process or driven equipment in accordance with the user program logic.

Allen-Bradley PLC-5 File Structure

The Allen-Bradley PLC-5 processors are usually programmed with an IBM® compatible computer and the areas of memory are referred to as files, not tables, as is the case with the PLC-2 memory structure. Although there are still two memory sections (storage [data] and user [program]), the PLC-5 memory map, or structure, is very flexible in the way that the memory can be allocated. Figure 5–12 shows the PLC-5 **default** memory structure. Default refers to the initial value, setting, or configuration prior to any user changes.

In the data or storge memory section file 0 is the output image file. This file has 32 words of 16 bits each, and can hold the status of 512 real-world output devices (32 × 16). The status of the outputs (*ON* or *OFF*) is updated once each scan. On some PLC-5 models, like the PLC-5/25, the size of the file can be increased to accommodate more output devices.

File 1 is the input image file. This file, like file 0, has 32 words of memory and can store the status of 512 input devices. The status (ON or OFF) of the input devices, like the output image file, is updated once each scan and can be increased in size on some PLC-5 models.

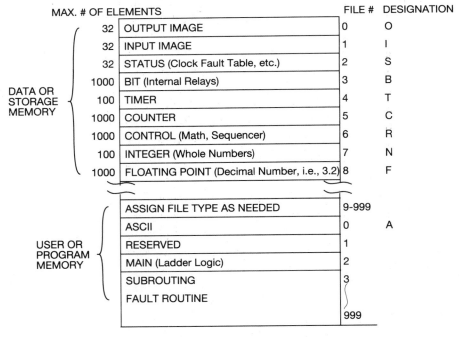

Figure 5–12 PLC-5 File Structure

Both files 0 and 1 use the octal numbering system, and the memory locations (bits) are also numbered using the octal numbering system (there are no 8s or 9s in the octal numbering system). The digits are 0–7, 10–17, 20–27, and so forth.

File 2 is the status, or S file. This file is used to store information on general processor status, fault codes, real time clock and calendar, major and minor fault bits, and program scan times in msecs. Information from this file is used or incorporated into the user program. The size of this file changes depending on the processor that is being used.

File 3 is the B, or bit file, and is used primarily for internal or dummy relays. The default size of this file is one word, but can be expanded to 1000 words if needed. All addresses from this file must start with B3. Another B or bit file may be created using the other areas of the memory. A B10 file could be created that would also have internal or dummy relays. The addressing B10 versus B3 is used for organization and ease of identification. The B3 file may be associated with one piece of equipment, while the B10 file could be associated with another piece of equipment and/or operation. Up to 999 files can be created in a PLC 5/15 processor memory, as long as you do not try to allocate more memory than is physically available in the processor. The B3 file is typical of the remaining files in flexibility, as well as being addressed using the decimal numbering system.

File 4 is the T, or timer file. All timer addresses must start with T4 unless new timer files have been created (i.e., T 9, T 10, and T 11). When creating files, the same number cannot be used twice. If file 10 is used as a B file (B10), then file 10 cannot be used as a timer file. Each timer that is programmed uses three words of memory from its timer file.

File 5 is the C, or counter file. All counters that are programmed have C5 as the start of their addresses. Each counter, like timers, use three words of the counter file memory.

File 6 is the R, or control file. The words in this file are used with special functions like sequencer, file moves, word to file moves, and math functions. File 7 stores whole numbers (integers) and is called the N, or integer file. The integer file is used to store numeric values for data compare, arithmetic functions, and the like. For storing numbers with a decimal point, or floating point, file 8, the F file is used. Files 9–999 can be used or assigned as needed. They may be used to expand the size of input or output files, timer and counter files, etc.

The program portion, or user portion, of memory (Figure 5–12) is used to store information that relates to the user program, or for information that is needed for the processor to operate. File 0 is used to store ASCII information, while File 1 is reserved for internal use by the processor. File 2 is where the user program is stored in RELAY LADDER LOGIC. Files 3–999 are for storing subroutines, fault routines, and selectable timed interrupt (STI) as they are needed.

Figure 5–13 shows the data table file structure used by the various PLC-5 models. Note that the I/O section varies from 32 words for the input image table and 32 words for the output image table for the PLC-5/10, 5/12, 5/15, 5/11, 5/20, and 5/20E, while there are 192 words for both the input and output image tables for the PLC-5/60, 5/60L, and 5/80.

Data Table Files

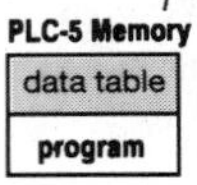

File Description		Number (Default File)	Maximum Size of File (16-bit words)						Memory Used Classic PLC-5 Processors	Memory Used Enhanced PLC-5 Processors
			PLC-5/10, -5/12, -5/15	PLC-5/11, 5/20, -5/20E	PLC-5/25	PLC-5/30	PLC-5/40, -5/40E, -5/40L	PLC-5/60, -5/60L, -5/80		
Output Image	O	0	32	32	64	64	128	192	2/file + 1/word	6/file + 1/word
Input Image	I	1	32	32	64	64	128	192	2/file + 1/word	6/file + 1/word
Status	S	2	32	128	32	128	128	128	2/file + 1/word	6/file + 1/word
Bit (binary)	B	3-999 (3)	1000						2/file + 1/word	6/file + 1/word
Timer	T	3-999 (4)	1000 structures of 3						2/file + 3/structure	6/file + 3/structure
Counter	C	3-999 (5)	1000 structures of 3						2/file + 3/structure	6/file + 3/structure
Control	R	3-999 (6)	1000 structures of 3						2/file + 3/structure	6/file + 3/structure
Integer	N	3-999 (7)	1000						2/file + 3/word	6/file + 3/word
Floating Point	F	3-999 (8)	1000						2/file + 2/float word	6/file + 2/float word
ASCII	A	3 - 999	1000						2/file + 1/2 per character	6/file + 1/2 per character
BCD	D	3 - 999	1000						2/file + 1/word	6/file + 1/word
Block Transfer[1]	BT	3 - 999	1000 structures of 6							6/file + 6/structure
Message[1]	MG	3 - 999	585 structures of 56							6/file + 56/structure
PID[1]	PD	3 - 999	399 structures of 82							6/file + 82/structure
SFC Status[1]	SC	3 - 999	1000 structures of 3							6/file + 3/structure
ASCI String[1]	ST	3 - 999	780 structures of 42							6/file + 42/structure
Extra Storage		3 - 999								

[1] Enhanced PLC-5 processors only.

PLC-5 Memory
  data table
  program

Figure 5–13 Data Table Map File Structure for the PLC-5 Family
(Courtesy of Allen-Bradley)

The second column of the chart shows the file numbers for the default files. Files 0, 1, and 2 are fixed and cannot be changed. Files 3–8, however, can be changed from the default settings and used as required. For example, if one wanted to use file 3—the binary file—for a timer file, it would be necessary to delete the binary file. The binary file is deleted from the data table map screen and then used as a timer file. Because files 3–8 can be changed, files 3–999 can then be used for timer files, counter files, and the like. However, it is much easier to use files 9–999 when additional files are needed.

While the names of files are different for each PLC manufacturer, the flexibility of assigning file areas and file size is typical for all PLCs. As the names and structure vary, the only way to really understand the memory structure is to obtain the literature for the specific PLC that you are dealing with. Salesmen, saleswomen and technical representatives are all invaluable resources when trying to gather information or clarification about a particular PLC.

Chapter Summary

All data, logic, and numerics are stored with binary digits that are represented as either a 1 or a 0. By storing binary data, the processor can rapidly scan and execute the user program and update the I/O section. I/O addresses in many cases not only identify the word and bit that is associated with the I/O, but also indicate hardware location (rack, module group, and terminal).

The names given to memory sections or subsections are unique to each PLC manufacturer, but the memories all work in basically the same manner.

The processor memory (storage and user) stores the I/O status, the user program, and numeric data used by the processor.

REVIEW QUESTIONS

1. The following types of information are normally found and/or stored in one of the PLC's two memory categories (user and storage). Place an S (for storage memory) or a U (for user memory) before the information type to indicate in which category it is normally found and/or stored.
   a. status of discrete input devices
   b. preset values of timers and counters
   c. numeric values of arithmetic
   d. holding registers
2. T   F   When a PLC is first turned *ON*, it will run a self-diagnostic or self-check test.
3. Describe the three steps of a typical PLC processor scan.
4. T   F   The actual scan time, or time it takes the PLC to complete a 3-step scan, decreases as the number of program words increases.
5. Identify the following PLC-5 files:
a. I	b. O	c. N
d. S	e. B	f. T
g. R	h. F	i. C
6. Define the term *byte*.
7. In a PLC-2, in what rack and module group number would address 11103 be located?
8. What word and bit number are represented by PLC-5 address O:010/01?
9. Using Square D Model 300 memory organization, would address 04-01 indicate an internal or external I/O?

Chapter 6 Numbering Systems

Objectives

After completing this chapter, you should have the knowledge to
- Understand decimal, binary, octal, hexadecimal, and binary coded decimal (BCD) numbering systems.
- Convert from one numbering system to another.

An electrician, technician, or other personnel who are required to program, modify, or maintain a PLC must have a "working" knowledge of the different numbering systems that are used. For example, the input/output addresses may use the octal numbering system; the timer and counter addresses may use the decimal numbering system; accumulated and preset values of the timers and counters may use the binary numbering system; operator interfaces, such as thumbwheels and seven segment displays, may require information be sent and received using the BCD format; and the hexadecimal system may be used for loading information into sequencers. The numbering system used in each area discussed varies with the different PLC manufacturers, but it is obvious that to fully understand and program a PLC, an understanding of the various numbering systems is necessary.

Decimal System

The decimal numbering system is used every day by electricians and technicians, and it is a system they are comfortable with. This system uses ten unique numbers, or digits, which are 0 through 9. A numbering system that uses 10 digits is said to have a base of 10. The value of the decimal number depends on the digit(s) used, and each's place value. Each position can be represented as a power of 10, starting with 10^0 as shown in Figure 6–1. In the decimal system, the first position to the left of the decimal point is called the **units place**, and any digit from 0–9 can be used. The next position to the left of the units place is the **tens place**; next is the **hundreds place**, the **thousands place**, and so on, with each place extending the capability of the decimal system by ten, or a power often.

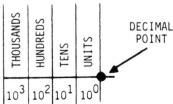

Figure 6–1 Place Value and Corresponding Power of Ten

NOTE: Any number that uses an exponent of 0, such as 10^0 has a place value of 1. Exponent 10^0 equals 1.

A specific decimal number can be expressed by adding the place values as shown in Figure 6–2.

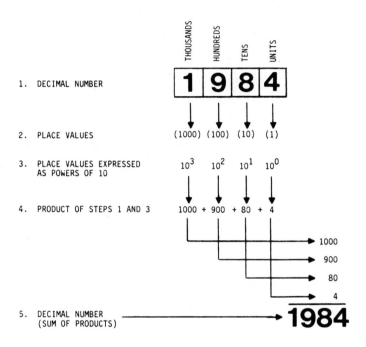

Figure 6–2 Decimal Numbering System

Mathematically, each place value is expressed as a digit number times a power of the base, or 10, in the decimal numbering system.

Another example is shown in Figure 6–3 using the decimal number 239.

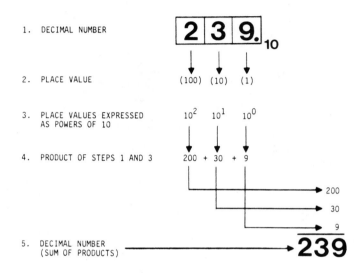

Figure 6–3 Decimal Numbering System

Binary System

The binary system uses only two digits: 1 and 0. Since only two digits are used, this system has a base of 2. Like the decimal system—and all numbering systems for that matter—each digit has a certain place value. The first place to the left of the starting point, or binary point, is the units or 1s location (base 2^0). The next place, to the left of the units place, is the 2s place, or base 2^1 as shown in Figure 6–4. The next place value is the 4s place, or base 2^2, then the 8s place, or base 2^3, and so forth. A binary number is always indicated by placing a 2 in subscript to the right of the units digit. Figure 6–4 illustrates how a binary number is converted to a decimal equivalent number. Note the subscripted 2 at the lower right-hand corner of the binary number line that indicates a base 2, or binary number.

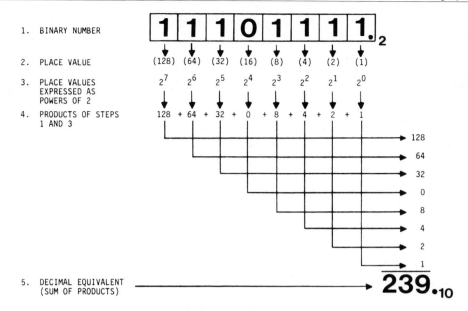

1.	BINARY NUMBER	1 1 1 0 1 1 1 1 ._2_
2.	PLACE VALUE	(128) (64) (32) (16) (8) (4) (2) (1)
3.	PLACE VALUES EXPRESSED AS POWERS OF 2	2^7 2^6 2^5 2^4 2^3 2^2 2^1 2^0
4.	PRODUCTS OF STEPS 1 AND 3	128 + 64 + 32 + 0 + 8 + 4 + 2 + 1

128
64
32
0
8
4
2
1

5. DECIMAL EQUIVALENT (SUM OF PRODUCTS) → **239.**₁₀

Figure 6–4 Converting a Binary Number to a Decimal Number

To convert a decimal number into a binary number, or to any numbering system for that matter, use the following procedure as shown in Figure 6–5. Divide the decimal number by the base you wish to convert to, in this case 2. The remainder is the 1s value (see Step 1 in the figure). Now divide the quotient from the first division again; the remainder becomes the value that is placed in the 2s location (see Step 2). The quotient of each preceding division is then divided by the base 2 until the base can no longer be divided (see Step 8), and the remainder (1) becomes the last digit in the binary number.

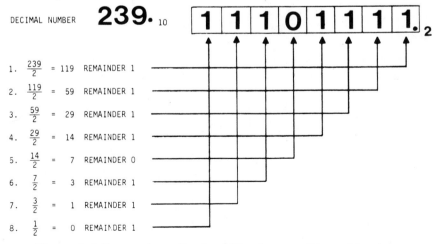

DECIMAL NUMBER **239.**₁₀   1 1 1 0 1 1 1 1 ._2_

1. $\frac{239}{2}$ = 119 REMAINDER 1
2. $\frac{119}{2}$ = 59 REMAINDER 1
3. $\frac{59}{2}$ = 29 REMAINDER 1
4. $\frac{29}{2}$ = 14 REMAINDER 1
5. $\frac{14}{2}$ = 7 REMAINDER 0
6. $\frac{7}{2}$ = 3 REMAINDER 1
7. $\frac{3}{2}$ = 1 REMAINDER 1
8. $\frac{1}{2}$ = 0 REMAINDER 1

Figure 6–5 Converting a Decimal Number to a Binary Number

It is important to arrange the remainders correctly when making the decimal-to-binary conversion. The first digit placed in the 1s position is called the *least* significant digit, whereas the last digit is called the *most* significant digit. The last digit placed has the highest place value (128s) which is why it is called the most significant digit. This reference to least and most significant digits is common, and refers to the relative position of any given digit within a number.

The following steps summarize this decimal-to-binary conversion.

Step 1.   The decimal number is divided by 2 (base of the binary numbering system). The quotient is listed (119) as well as the remainder (1).

Step 2.   Divide the quotient of Step 1 (119) by base 2, and list the new quotient (59) and the remainder (1).

Step 3.   Divide the quotient of Step 2 (59) by base 2, and list the new quotient (29) and remainder (1).

Step 4.   Divide the quotient of Step 3 (29) by 2, and list the new quotient (14) and the remainder (1).

Step 5.   Divide the quotient of Step 4 (14) by 2, and list the new quotient (7) and remainder (0).

Step 6.   Divide the quotient of Step 5 (7) by 2, and list the new quotient (3) and remainder (1).

Step 7.   Divide the quotient of Step 6 (3) by 2, and list the new quotient (1) and remainder (1).

Step 8.   Divide the quotient of Step 7 (1) by 2, and list the new quotient (0) and remainder (1).

NOTE: When using a calculator to do the division, the value to the *right* of the decimal must be multiplied by the base to get the actual remainder. For example, when 239 is divided by 2 (Step 1) on a calculator, the answer is 119.5. To find the actual remainder, the 0.5 is multiplied by 2, the base, to find the remainder 1. This procedure is true for any numbering system. The base times the value to the right of the decimal point equals the actual remainder.

The binary numbering system is used to store information in the processor memory in the form of *bits* (BInary digiTS).

Octal System

The octal system, or base 8, is made up of eight digits: numbers 0 through 7. The first digit to the *left* of the octal point is the units place, or 1, and has a base or power of 8^0. The next place is eights (8s) or base 8^1. The next place is sixty-fours (64s) or base 8^2, followed by five hundred twelves (512s) or base 8^3, and four thousand ninety-sixes or base 8^4, and so on. An octal number will always be expressed by placing an eight in subscript to the right of the units digit as shown in Figure 6–6.

$$357._8$$

Figure 6–6 Octal Number

The method of converting an octal number to a decimal equivalent number is illustrated in Figure 6–7.

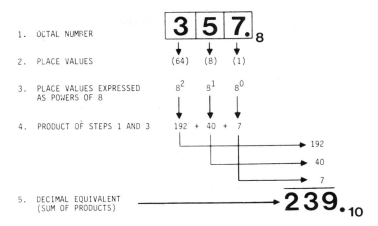

Figure 6–7 Converting an Octal Number to a Decimal Number

The decimal number 239 is converted to an octal number in Figure 6–8.

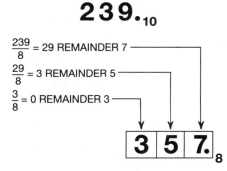

Figure 6–8 Converting a Decimal
Number to an Octal Number

Step 1.   The decimal number 239 is divided by 8 (base for the octal numbering system). The quotient is listed (29) as well as the remainder (7). A calculator shows the answer as 29.875. The quotient is 29, and the remainder is 0.875×8, or 7.

Step 2.   Divide the quotient of Step 1 (29) by 8, and list the new quotient (3) and the remainder (5). A calculator gives the answer 3.625. The quotient is 3, and the remainder is 0.625×8, or 5.

Step 3.   Divide the quotient from Step 2 (3) by 8, and list the new quotient (0) and remainder (3). The quotient 3 divided by 8 equals 0.375. The new quotient is 0, and the remainder is 0.375×8, or 3.

The decimal number 239 is the same as octal number 357.

Since the largest single number that can be expressed using the octal numbering system is seven (7), each octal digit can be represented by using only three (3) binary bits (base 2). Figure 6–9 illustrates how to convert an octal number to a binary number. The figure shows three sets of binary bits and the place value of each bit. For the *least* significant digit (7), a one (1) must be placed in the 1s place, the 2s place, and the 4s place to equal 7. The middle digit (5), a 1, is placed in the 1s place and the 4s place, while a 0 is placed in the 2s place. This combination equals 5. For the *most* significant digit (3), a 1 is placed in the 1s place and the 2s place, and a 0 is placed in the 4s place. This combination adds up to 3.

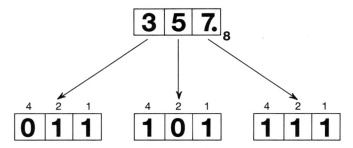

Figure 6–9 Conversion of Octal Number to Binary

Allen-Bradley uses the octal numbering system for I/O addressing in the PLC-2 and PLC-5 families. The terminals of the input and output modules are labeled 00 through 07 and 10 through 17, rather than 0 through 15 as would be the case with decimal numbering. The Allen-Bradley PLC-2 also uses the octal numbering system for numbering words and bits. When using the octal numbering system, words are labeled 000–007, 010–017, 020–027, and so forth, whereas the bits are labeled 00–07 and 10–17. Figure 6–10 shows a memory word with the internal bits addressed using the octal numbering system.

17	16	15	14	13	12	11	10	07	06	05	04	03	02	01	00

WORD 010

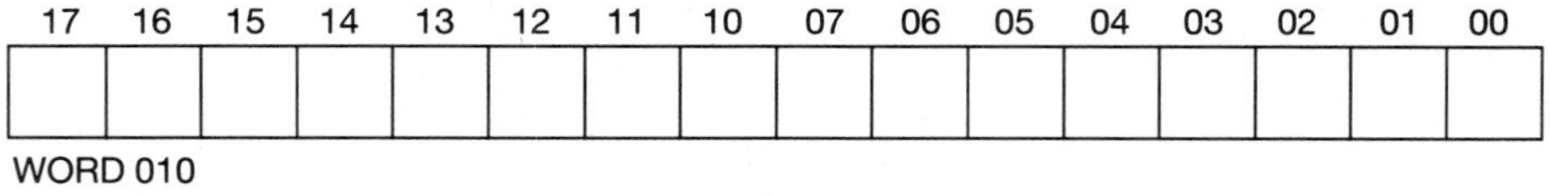

Figure 6–10 Word and Bit Labeling Using the Octal Numbering System

Hexadecimal System

The hexadecimal system, often referred to as HEX, consists of a number system with base 16.

It seems logical that the numbers used in base 16 would be 0 through 15. However, only numbers 0 through 9 are used, and the letters A through F represent numbers 10–15, respectively. The place values from the hexadecimal point are 1s—16^0, 16s—16^1, 256s—16^2, 4096s—16^3, and so on.

Each hexadecimal digit is represented by four (4) binary digits. The binary equivalents are shown in the table in Figure 6–11.

HEXADECIMAL	BINARY	DECIMAL
0	0000	0
1	0001	1
2	0010	2
3	0011	3
4	0100	4
5	0101	5
6	0110	6
7	0111	7
8	1000	8
9	1001	9
A	1010	10
B	1011	11
C	1100	12
D	1101	13
E	1110	14
F	1111	15

Figure 6–11 Hexadecimal Equivalents for Binary and Decimal

The decimal number 4,780 is converted to hexadecimal as illustrated in Figure 6–12.

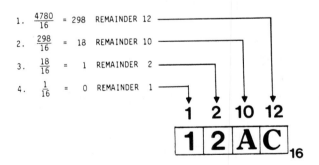

DECIMAL NUMBER $4780._{10}$

1. $\frac{4780}{16}$ = 298 REMAINDER 12
2. $\frac{298}{16}$ = 18 REMAINDER 10
3. $\frac{18}{16}$ = 1 REMAINDER 2
4. $\frac{1}{16}$ = 0 REMAINDER 1

1 2 10 12

$12AC_{16}$

Figure 6–12 Converting a Decimal Number to a Hexadecimal Number

Step 1. The decimal number is divided by 16 (base for the hexadecimal numbering system). The quotient is listed (298) as well as the remainder (12). A calculator provides the answer 298.75. The quotient is 298, and the remainder is 0.75 × 16, or 12.

Step 2. Divide the quotient of Step 1 (298) by 16, and list the new quotient (18) and the remainder (10). The answer is 18.625. The quotient is 18, and the remainder is 0.625 × 16, or 10.

Step 3.   Divide the quotient from Step 2 (18) by 16, and list the new quotient (1) and the remainder (2). Eighteen divided by 16 equals 1.125. The quotient is 1, and the remainder is 0.125 × 16, or 2.

Step 4.   Divide the quotient from Step 3 (1) by 16, and list the new quotient (0) and the remainder (1). One divided by 16 equals 0.0625. The quotient is 0, and the remainder is 0.0625 × 16, or 1.

Converting a hexadecimal number to a decimal number is illustrated in Figure 6–13.

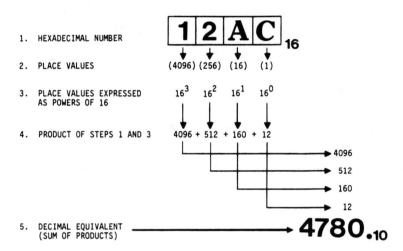

1. HEXADECIMAL NUMBER

2. PLACE VALUES

3. PLACE VALUES EXPRESSED AS POWERS OF 16

4. PRODUCT OF STEPS 1 AND 3

5. DECIMAL EQUIVALENT (SUM OF PRODUCTS)

Figure 6–13 Converting a Hexadecimal Number to a Decimal Number

NOTE: Remember that A is equivalent to 10, and C is equivalent to 12.

The binary equivalent of the hexadecimal number 12AC is shown in Figure 6–14.

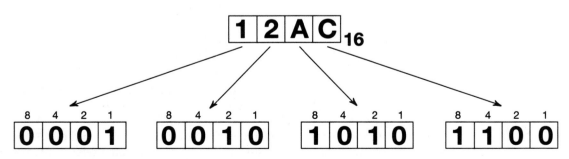

Figure 6–14 Binary Equivalent of a Hexadecimal Number

Since the largest number that can be displayed using the hexadecimal numbering system is 15, or F (as shown in the table in Figure 6–11), only four binary bits are needed to display each hexadecimal digit. The conversion to binary simply places the 1s in the correct binary locations to duplicate the hexadecimal digit (1 through F), as illustrated. The C has a value of 12, so a 1 is placed in the 8s and 4s location, while zeros (0) are placed in the 2s and 1s location for a total binary value of 12. The same procedure is followed for the remaining digits A (10), 2, and 1.

The HEX system is used when large numbers need to be processed. The hexadecimal system is also used by some PLCs for entering output instructions into a sequencer.

Binary Coded Decimal (BCD) System

When large decimal numbers are to be converted to binary for memory storage, the process becomes somewhat cumbersome. To solve this problem and speed conversion, the Binary Coded Decimal (BCD) system was devised. In the BCD system, four binary digits (base 2) are used to represent each decimal digit. To distinguish the BCD numbering system from a binary system, the designation BCD is subscripted and placed to the lower right of the units place. Converting a BCD number to a decimal equivalent is shown in Figure 6–15.

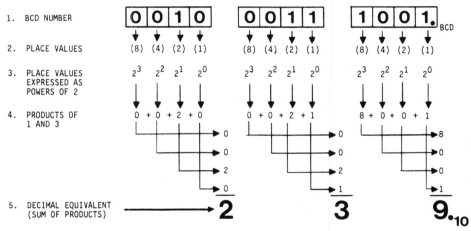

Figure 6–15 Converting a BCD Number to a Decimal Number

When using a BCD numbering system, three decimal numbers may be displayed using 12 bits (3 groups of 4), or 16 bits (4 groups of 4) may be used to represent four decimal numbers or digits.

When only three decimal digits are to be represented, using 12 bits, they are further identified as *most significant digit* (MSD), *middle digit* (MD), and *least significant digit* (LSD) (Figure 6–16).

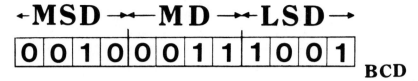

Figure 6–16 Identifying BCD Digits

When using the BCD system, the largest decimal number that can be displayed by any four binary digits is 9. The table in Figure 6–17 shows the four binary digit equivalents for each decimal number 0 through 9.

PLACE VALUE				DECIMAL EQUIVALENT
2^3 (8)	2^2 (4)	2^1 (2)	2^0 (1)	
0	0	0	0	0
0	0	0	1	1
0	0	1	0	2
0	0	1	1	3
0	1	0	0	4
0	1	0	1	5
0	1	1	0	6
0	1	1	1	7
1	0	0	0	8
1	0	0	1	9

Figure 6–17 Binary to Decimal Equivalents

Using Numbering Systems

The keyboards used by various types of programming devices may have electric symbols and special function keys, along with numeric, or number keys, for addressing. Many keyboards also have alphanumeric (letters and numbers) keys for report generation and other special programming functions.

The alphanumeric keys of many programming terminals generate standard ASCII characters and control codes. **ASCII** is an acronym for **A**merican **S**tandard **C**ode for **I**nformation **I**nterchange. The ASCII code uses different combinations of 7 bit binary (base 2) information for communication of data. The data may be communicated to a printer, tape loader, floppy or hard disc, or be displayed on the VDT of the programmer and/or computer.

NOTE: ASCII information is often expressed in hexadecimal (base 16.) Figure 6–18 shows the 128 standard ASCII control code and character set with both the binary and hexadecimal numbering systems.

MSB
MOST SIGNIFICANT BIT

	BINARY→	000	001	010	011	100	101	110	111
	HEX→	0	1	2	3	4	5	6	7
000	0	NUL	DLE	SP	Ø	@	P	\	p
0001	1	SOH	DC1	!	1	A	q	a	q
0010	2	STX	DC2	"	2	B	R	b	r
0011	3	ETX	DC3	#	3	C	S	c	s
0100	4	EOT	DC4	$	4	D	T	d	t
0101	5	ENQ	NAK	%	5	E	U	e	u
0110	6	ACK	SYN	&	6	F	V	f	v
0111	7	BEL	ETB	'	7	G	W	g	w
1000	8	BS	CAN	(	8	H	X	h	x
1001	9	HT	EM	)	9	I	Y	i	y
1010	A	LF	SUB	*	:	J	Z	j	z
1011	B	VT	ESC	+	;	K	[	k	{
1100	C	FF	FS	'	<	L	\	l	¦
1101	D	CR	GS	-	=	M	]	m	}
1110	E	SO	RS	.	>	N	^	n	~
1111	F	SI	US	/	?	O	–	o	DEL

(leftmost label, vertical: LSB / LEAST SIGNIFICANT BIT)

Figure 6–18 Standard ASCII Control Code and Character Set

The digital or hexadecimal number is determined by first locating the vertical column where the code or character is located, and then the horizontal line.

EXAMPLE: The letter A is in column 4, horizontal line 1. The binary number that transmits the letter A is 100 0001. The hexadecimal number is 41. The symbol # is 010 0011 in binary and 23 in HEX.

An eighth bit is often used by programmers to provide error checking of information that is transmitted. This eighth bit is called the **parity bit**.

For *even* parity, the parity bit (the eighth bit) is added to the seven bits that represent the ASCII codes and characters so that the number of 1s will always add up to an even number.

EXAMPLE: The binary number for the # symbol is 010 0011. The 1s add up to three, an odd number. By adding an eighth bit and making it a 1, the total of 1s are now 4, or even, as shown in Figure 6–19.

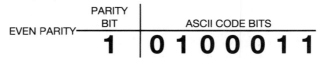

Figure 6–19 Parity Bit Set to 1 for Even Parity

The letter A, which is the binary number 100 0001, has two 1s and is already even. In this case, the parity bit would be a 0, as shown in Figure 6–20.

Figure 6–20 Parity Bit Set to 0 for Even Parity

The ASCII control code BS (backspace) is binary number 000 1000. For even parity, a 1 is added for the parity bit shown in Figure 6–21.

Figure 6–21 Even Parity

By checking each character or control code that is sent for an even number of 1s, transmission errors can be detected when an odd number of 1s is found.

For systems that operate on *odd* parity, the parity bit is used to make the total of 1s add up to an odd number.

EXAMPLE: The number 5 has a binary number of 011 0101. The 1s add up to 4. The parity bit is set to 1, making the 1s total 5, or an odd number. Figure 6–22 illustrates this concept.

Figure 6–22 Parity Bit Set to 1 for Odd Parity

For systems that do not use a parity bit for error checking, the eighth bit is always a zero (0).

Chapter Summary

There are several numbering systems that are used to store information in the form of binary digits (bits) into the memory system of a processor. The specific numbering system or the combination of numbering systems used depends on the hardware requirements of the specific PLC manufacturer. The important thing to remember, however, is that no matter which numbering system or systems are used, the information is still stored as 1s and 0s.

REVIEW QUESTIONS

1. When information is stored using only 1s and 0s, it is called a _____ system.
2. A *bit* is an acronym for _____ .
3. The decimal numbering system uses 10 digits, or a base of 10. List the base for each of the following numbering systems.
   a. binary   base _____
   b. hexadecimal   base _____
   c. octal   base _____
4. Convert *binary* number 11011011 to a *decimal* number.
5. Convert *decimal* number 359 to a *binary* number.
6. Convert *hexadecimal* number 14CD to a *decimal* number.
7. Convert *decimal* number 3247 to a *hexadecimal* number.
8. Convert *decimal* number 232 to an *octal* number.
9. How do we prevent binary numbers 10 and 11 from being confused as decimal numbers?
10. Convert the following *binary* values to *decimal*.
    a. 10011000
    b. 01100101
    c. 10011001
    d. 00010101
11. Convert the following *BCD* values to *decimal*.
    a. 1001 1000
    b. 0110 0101
    c. 1001 1001
    d. 0001 0101
12. The BCD value 1001 0011 0101 is *not*
    a. 935 decimal
    b. 0011 1010 0111 binary
    c. 647 octal
    d. 3A7 hexadecimal
13. The hexadecimal value 2CB is *not*
    a. 715 decimal
    b. 1313 octal
    c. 0010 1100 1011 binary
    d. 0111 0001 0011 BCD

Chapter 7

Understanding and Using Ladder Diagrams

Objectives

After completing this chapter, you should have the knowledge to

- Identify a wiring diagram.
- Identify the parts of a wiring diagram.
- Convert a wiring diagram to a ladder diagram.
- List the rules that govern a ladder diagram.

There are basically two types of electrical diagrams: wiring diagrams and ladder diagrams.

Wiring Diagrams

The wiring diagram shows the circuit wiring and its associated devices (relays, timers, motor starters, switches, and the like) in their relative physical location (Figure 7–1). While this type of diagram assists in locating components and shows how a circuit is actually wired, it does not show the circuit in its simplest form. To simplify understanding of how a circuit works, and to show the electrical relationship of the components (not the physical relationship), a ladder diagram is used.

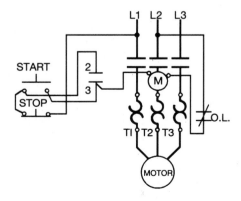

Figure 7–1 Wiring Diagram

Ladder Diagrams

The ladder diagram, also referred to as a schematic or elementary diagram, is used by the electrician or technician to speed their understanding of how a circuit works. Figure 7–2 shows the same circuit as Figure 7–1, but in ladder diagram form.

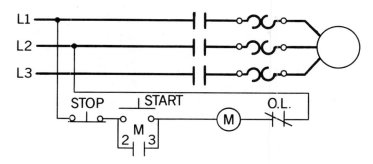

Figure 7–2 Ladder Diagram

To simplify the circuit and help to understand its configuration, the power portion of the circuit is shown separate from the control portion. No attempt is made to show the actual physical location of the components. Since the motor connections (power portion) are the same for any three-phase motor, it is common practice not to show the motor starter or the motor. By not showing the power portion of the circuit, a ladder diagram is created, showing only the control portion of the diagram (Figure 7–3).

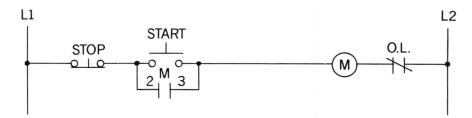

Figure 7–3 Simplified Ladder Diagram

The power required for the control circuit is always shown as two vertical lines, while the actual line(s) of logic are drawn as horizontal lines. The power lines, or rails as they are often called, are like vertical sides of a ladder, whereas the horizontal logic lines are like the Rungs of a ladder.

When referring back to Figure 7–1, it is easy to see the physical relationship between the *STOP/START* station, the motor starter coil (M), the overload contacts (O.L.), and the holding contacts (2 and 3), but it is difficult to determine the electrical relationship. The ladder diagram in Figure 7–3, however, clearly shows the electrical relationship between all of the control circuit components.

Ladder Diagram Rules

Some basic rules for ladder diagrams are as follows:

1. A ladder diagram is read like a book; from left to right and from top to bottom.
2. The vertical power lines (rails) of the ladder diagram represent the *voltage potential* of the circuit. The potential could be AC or DC, and varies in voltage from 6 V to 480 V. Standard labeling for the rails is L1 and L2. L1 is AC high or hot for AC circuits, and positive or plus (+) for DC circuits. L2 is AC low or neutral for grounded AC circuits, and negative or minus (−) for DC circuits. The rails may also be marked X1 and X2 when the voltage potential is derived from a transformer.
3. Devices or components are shown in order of importance whenever possible. In Figure 7–3 the *STOP* button is shown ahead of the *START* button. For safety reasons, the *STOP* button has a higher order of importance than the *START* button.
4. Electrical devices or components are shown in their normal condition. The normal condition of electrical diagrams is the circuit deenergized (*OFF*) and with no external forces such as pressure, or flow, etc., acting on the device. The *STOP* button is shown closed because that is the normal position for the *STOP* button. The holding contacts (2 and 3) of coil M are shown open. This is the normal position for these contacts when coil M is deenergized. The normally open (N.O.) M holding contacts 2 and 3 do not close until there is a complete path for current flow to coil M. When coil M energizes, M contacts 2 and 3 close, providing parallel path for current flow with the *START* button.
5. Contacts associated with relays, timers, motor starters, and the like, always have the same number or letter designation as the device that controls them. This labeling method holds true no matter where the contacts(s) appear in the circuit. For example, in Figure 7–3 the N.O. holding contacts 2 and 3 are controlled (activated) by motor starter coil M. Therefore, the contacts are identified with the letter M.
6. All contacts associated with a device change position when the device is energized.

   Figure 7–4 shows a control relay (CR) controlled by a switch (S-1) on Rung 1 of the ladder diagram. Rung 2 shows a normally closed (N.C.) control relay contact in series with a green indicator lamp. Rung 3 shows a normally open (N.O.) control relay contact in series with a red indicator light.

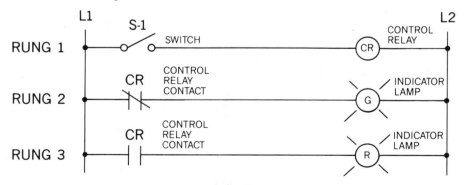

Figure 7–4 Three Rung Ladder Diagram

When power is applied to the rails of the ladder diagram, the only device in the circuit that operates is the green indicator lamp. The green indicator lamp lights due to a complete path for current flow through the normally closed (N.C.) control relay contacts. These contacts are normally closed and only change position and open when the control relay in Rung 1 is energized. When switch S-1 is closed, completing the path for current flow and energizing the CR in Rung 1, the N.C. CR contacts in Rung 2 open, while the N.O. CR contacts in Rung 3 close. The action of the contacts will turn *OFF* the green lamp in Rung 2 and turn *ON* the red lamp in Rung 3. As long as the control relay remains energized through S-1, the normally closed contact in Rung 2 remains open, and the normally open contact in Rung 3 remains closed. When S-1 is opened and CR deenergizes, the contacts controlled by CR will return to their normal state (N.C. in Rung 2 and N.O. in Rung 3).

7. In a ladder diagram, devices that perform a *STOP* function are normally wired in series. Figure 7–5 shows two switches wired normally closed (N.C.) that control a green indicator lamp.

Figure 7–5 Two Switches Wired in Series

With the two switches wired in series, both A and B must remain closed for the lamp to remain lit. If either switch is opened, the green lamp will go out. When switches and/or contacts are wired in series, they are said to have an AND relationship. The AND relationship requires that both A *and* B must be closed for the lamp to light. A truth table for this concept is shown in Figure 7–6.

SWITCH	SWITCH	INDICATOR LAMP
A	B	G
OFF	OFF	OFF
OFF	ON	OFF
ON	OFF	OFF
ON	ON	ON

Figure 7–6 Truth Table for Series Devices

8. Devices that perform a *START* function are normally wired in parallel. Figure 7–7 shows two switches (A and B) wired in parallel to control a red indicator lamp. In this configuration, if either switch A *or* B is closed, the red lamp will light.

Figure 7–7 Two Switches Wired in Parallel

When switches or contacts are wired in parallel, they are said to have an OR relationship. The OR relationship requires that either A *or* B be closed for the red indicator lamp to light. A truth table for this concept is shown in Figure 7–8.

SWITCH	SWITCH	INDICATOR LAMP
A	B	R
OFF	OFF	OFF
OFF	ON	ON
ON	OFF	ON
ON	ON	ON

Figure 7–8 Truth Table for Parallel Devices

With this understanding of what a ladder diagram is, and the rules that apply to it, a discussion of a basic motor *STOP/START* circuit (shown in Figure 7–2) can begin.

Basic *STOP/START* Circuit

As stated earlier in this chapter, the wiring diagram in Figure 7–1 is great for showing actual physical location of the circuit wiring and the components. It does not, however, show the electrical relationship of the devices as simply as the ladder diagram. The wiring diagram is used for original installation and some troubleshooting, whereas the ladder diagram is used to show the electrical relationship of the components, and to speed understanding of how the circuit works.

From viewing the ladder diagram in Figure 7–9, it can be seen that when power is applied to the circuit, the motor starter coil M cannot energize because there is an incomplete path for current flow due to the open *START* button and the normally open M contacts (2 and 3). The *START* button and the N.O. M contacts are wired in parallel and have an OR relationship. When the *START* button is pushed, a path for current exists from L1 potential through the normally closed *STOP* button, through the now closed *START* button through the coil of the motor starter (M), and on through the N.C. overload contacts to L2 potential.

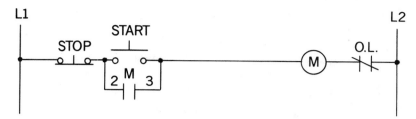

Figure 7–9 Ladder Diagram for Basic *STOP/START* Circuit

When the starter coil M energizes, the M contacts (2 and 3) close, providing an alternate path for current flow. At this point, the *START* button could be released, and the circuit would remain energized, or held in, by the holding contacts (2 and 3) of the motor starter. When contacts from a motor starter or other device are wired in this fashion, they are often referred to as **holding**, **maintaining**, or **sealing** contacts as the circuit is held, maintained, or sealed in after the *START* button is released.

When the holding contacts (2 and 3) are closed, the main motor contacts of the motor starter are also closed and the motor is started. The operation of the motor is normally taken for granted and is not shown on the ladder diagram. By keeping the ladder diagram as simple and uncluttered as possible, the relationship between components and how the control portion of the circuit works is greatly enhanced.

Figure 7–10 again shows the wiring diagram of a motor *STOP/START* circuit. While this diagram looks entirely different from the ladder diagram, both are electrically the same. This comparison shows the electrician or technician why the ladder diagram is preferred.

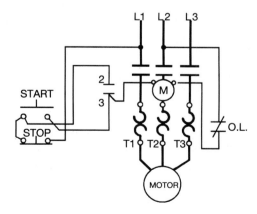

Figiure 7–10 Wiring Diagram for Basic *STOP/START* Circuit

The ladder diagram has been the "working language" of electricians and electrical engineers for many years, and helps explain why most programmable controllers are programmed using ladder logic. While this method of programming is welcomed by some, it has frustrated other PLC users who have not been exposed to, or trained in, RELAY LADDER LOGIC.

Sequenced Motor Starting

Relay ladder diagrams can become large and complex. It is not the purpose of this text to cover them in great detail, but instead to discuss the basic rules and present some concepts to enhance understanding of circuits that are discussed in later chapters.

Figure 7–11 shows a ladder diagram for a circuit that starts three motors.

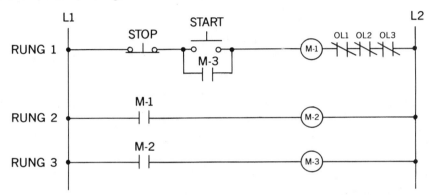

Firgure 7–11 Three Motor Start Circuit

Rung 1 contains the *STOP/START* buttons and the motor starter coil M-1 for Motor 1. Notice that the holding contacts wired in parallel with the *START* button are not M-1 contacts, but instead are M-3 contacts. With this arrangement, Rung 1 cannot be sealed in, or maintained, unless Motor Starter 3 energizes and closes its contacts. Additionally, the M-1 contacts in Rung 2 must close to energize Motor Starter 2 (M-2) and M-2 contacts in turn must close in Rung 3 to energize Motor Starter 3 (M-3). When the *START* button of this circuit is pushed, it operates as follows:

1. M-1 energizes, closing the N.O. M-1 contacts in Rung 2 and energizes M-2.
2. M-2 N.O. contacts close in Rung 3 and energize M-3.
3. M-3 N.O. contacts in Rung 1 close and act as holding contacts to keep the circuit energized after the *START* button is released.

NOTE: This sequence happens almost instantaneously.

4. Pushing the *STOP* button deenergizes M-1 which deenergizes M-2 in Rung 2 when the normally open M-1 contacts go open. M-2's deenergizing opens the M-2 contacts in Rung 3 and deenergizes M-3. With M-3 deenergized, the N.O. M-3 contacts in Rung 1 open. By wiring all three overload contacts in series with M-1 in Rung 1, an overload on any motor

would shut down all motors. An open overload contact would have the same effect as pushing the *STOP* button.

It could be said that this circuit consists of basically three elements: inputs, outputs, and logic.

The inputs consist of the *STOP* button, the *START* button, and the overload contacts. The outputs are motor starters M-1, M-2, and M-3. The logic that caused the sequential starting were normally open contacts M-1, M-2, and M-3.

These three elements—inputs, outputs, and logic—also work well with programmable controllers. The inputs are wired to input modules, the outputs are wired to output modules, and the processor performs the logic functions.

Figure 7–12 shows the wiring diagram for the three-motor circuit just discussed. This diagram further illustrates the point, that while wiring diagrams are great for giving the physical location of components, they do not show the control function of the circuit as clearly as a ladder diagram does.

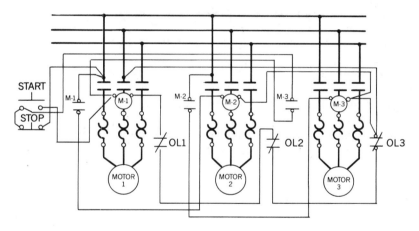

Figure 7–12 Wiring Diagram for Three Motor Circuit

Chapter Summary

There are basically two types of electrical diagrams: wiring diagrams and ladder diagrams. Wiring diagrams show actual physical location and wiring, whereas ladder diagrams show electrical relationships. The simplified ladder diagram speeds understanding of circuit operation and is used for circuit design and troubleshooting. The vertical sides of the ladder diagram are referred to as *rails*, while the horizontal lines or logic are called *Rungs*. On electrical diagrams, devices are always shown in their normal or deenergized condition. When two or more devices are wired in series, they perform an AND function, while two or more devices wired in parallel perform an OR function. The elements of the ladder diagram are inputs, outputs, and logic.

REVIEW QUESTIONS

1. Define the terms *normally open* and *normally closed*.
2. Describe the difference between a wiring diagram and a ladder (schematic) diagram.
3. Explain the operation of the circuit in Figure 7–9 if M contacts 2 and 3 do not close.
4. Contacts wired in parallel have what relationship?
   a. AND
   b. OR
5. Contacts wired in series have what relationship?
   a. AND
   b. OR
6. The two main vertical lines of a ladder diagram are often referred to as:
   a. Rungs
   b. power ports
   c. rails
   d. tracks
   e. none of the above
7. The horizontal lines of a ladder diagram are referred to as:
   a. Rungs
   b. power ports
   c. rails
   d. tracks
   e. none of the above
8. Devices that are intended to perform a *STOP* function are normally wired in_____with each other.
9. Devices that are intended to perform a *START* function are normally wired in_____with each other.
10. How are contacts that are associated with relays, motor starters, timers, and the like, identified?
11. Convert wiring diagram 7–A into a ladder diagram.

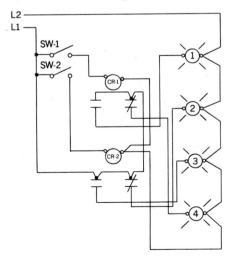

12. Convert wiring diagram 7–B into a ladder diagram.

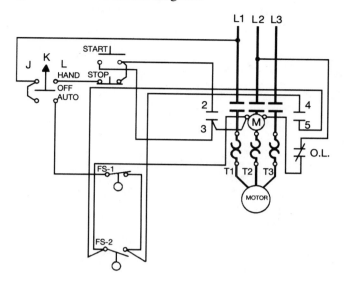

8

Relay Type Instructions

Objectives

After completing this chapter, you should have the knowledge to
- Understand the EXAMINE ON instruction.
- Understand the EXAMINE OFF instruction.
- Write and understand the logic for a standard *STOP/START* motor circuit.

The next step in understanding how the programmable logic controller works is to learn how ladder logic is changed into processor logic. The actual programming is accomplished using either a desktop, hand-held programming device, or computer. Figures 8–1a shows a Modicon Micro PLC and the Modicon Hand–Held Programmer for programming and monitoring. Figure 8–1b shows the keyboard of the hand-held programmer.

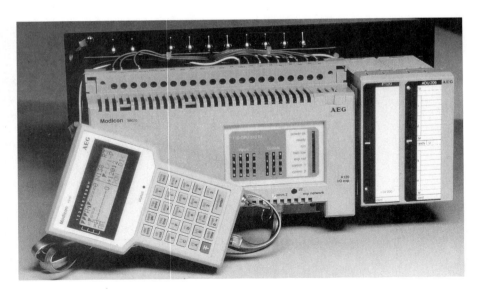

Figure 8–1a Modicon Micro and Hand-Held Programmer
(Courtesy of Modicon Inc.)

Regardless of the type of programmer used, some common relay symbols are standard. These symbols include normally open contacts, normally closed contacts, and coil or output. Figure 8–1b shows that the keyboard has no symbols for input devices such as *STOP* buttons, limit switches, and pressure switches. The contacts from all input devices are programmed using either the N.O. or N.C. relay contact symbols (above the letters A and B on the keyboard). The actual programming devices used by the different PLC manufacturers are covered in Chapter 9, but first the relay logic used for PLCs must be discussed and *understood*.

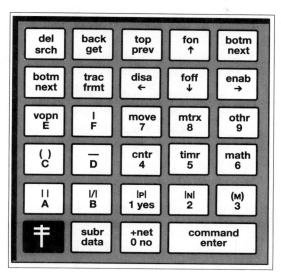

Figure 8–1b Modicon Hand-Held Programmer Keyboard

Programming Contacts

A PLC is normally programmed using a ladder logic-type language. Ladder logic is a good choice for a programming language because it closely resembles the way circuits are hard-wired. Electricians and technicians feel comfortable with, and understand, ladder logic, so programming with a ladder logic-type language makes good sense. While there are many similarities between standard RELAY LADDER LOGIC and the ladder logic used for programming a PLC, there are some distinct differences.

Hard-wired contacts in a motor control circuit control the path for current flow to the output (coil, light, solenoid, etc.). The contact symbols that are used when programming a PLC are actually logic instructions that the processor uses to make decisions.

The PLC normally open (N.O.) contact symbol is actually an instruction that tells the processor to look for an *ON* condition at the address that corresponds to the symbol. If an *ON* condition exists, the instruction is said to be logically *true*, and logic continuity exists. This is much like saying that if a contact is closed. current can flow.

NOTE: As stated earlier in the text, the terms current flow and power flow are often used by the various PLC manufacturers to indicate that a circuit is complete, or logically true.

Figure 8–2 shows a simple circuit containing a single pole switch and a lamp for an output. Figure 8–3 shows the equivalent circuit when programmed with a PLC. Addresses shown are Square D format. An "I" preceding a word and bit number indicates an input, whereas an "O" preceding a word and bit address indicates an output.

Figure 8–2 Simple Circuit

Figure 8–3 Equivalent Circuit Programmed with a PLC

In Figure 8–2, if the switch is closed, current flows through the switch contacts and the lamp lights. In Figure 8–3, the single-pole switch is shown as a normally open contact symbol with the address I01-01. Refering back to the memory structure in Chapter 5, the address I01-01 is bit 01 of I/O image word 01 using Square D format. The lamp, or output, is shown as a circle, and is given the address O01-09. The output is actually bit 09 of word 01.

Because of the way this program is written, the normally open contact symbol tells the processor to look at address location I01-01 (the single-pole switch); if a closed (*ON*) condition is found, then the logic of the circuit is true. When the logic of the circuit is true, the processor is instructed to turn *ON* output O10-09. There is, in reality, no actual electrical connection between the switch (I01-01) and the lamp (O01-09). Instead, it is the processor that turns the lamp *ON* or *OFF* depending on the logic of the program that is written and the status of the input device.

Figure 8–4 shows the switch and lamp as they are wired to their respective I/O modules. It is the N.O. contact symbol in the program that tells the processor to examine the single-pole switch for an *ON* condition. If the switch is open (*OFF*), the program logic is not true and the processor will not turn on the lamp. On the next processor scan, however, if the switch has been closed, it will be *ON*, the logic of the circuit will be true, and the processor will turn *ON* the lamp.

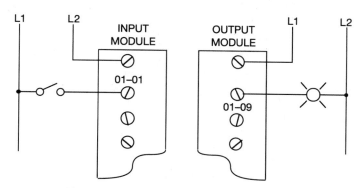

Figure 8–4 Input and Output Devices Wired to I/O Modules

Figure 8–5a shows the bit status of I/O word 1 when the switch is open. With the switch open (*OFF*), the logic of the circuit cannot be true so the lamp will not be *ON*. The *OFF* condition of the switch and the lamp is indicated by a 0 in bit location 01 and 09. When the switch is closed, the bit that represents the switch (01) will change to a 1. This makes the circuit logically true, the processor will turn the lamp to ON (bit 09), and a 1 is shown in that location (Figure 8–5b).

16	15	14	13	12	11	10	9	8	7	6	5	4	3	2	1
0	0	0	0	0	0	0	0	0	0	0	0	0	0	0	0

WORD 1

Figure 8–5a Bit Status for I/O Word 1 with Switch Open

16	15	14	13	12	11	10	9	8	7	6	5	4	3	2	1
0	0	0	0	0	0	0	1	0	0	0	0	0	0	0	1

WORD 1

Figure 8–5b Bit Status for I/O Word After Switch is Closed

As the processor views the normally open contact instruction as a request to examine a given address for an *ON* condition, it is referred to an EXAMINE ON instruction. The opposite instruction, the normally closed contact, is referred to as an EXAMINE OFF instruction.

The EXAMINE OFF instruction is only logically true when the device referenced is *OFF*, or open. Figure 8–6 shows the EXAMINE OFF instruction now used for address I01-01. The processor is asked to examine the address location for an *OFF* (open) condition. If an *OFF* condition is found, then the instruction is logically true and the output O01-09 would be turned *ON*. If, on the other hand, the switch was found to be *ON* (closed), the logic would be false and the lamp would not be turned *ON*.

Figure 8–6 Examine OFF Instruction

At first these two instructions may seem to be contrary to the logic of hard-wired contacts, so it is important to remember that these are instructions to the processor, and are not hard-wired contacts. A review of both instructions seems appropriate.

EXAMINE ON
Whenever the processor sees an N.O. contact in the user program, it views the contact symbol as a request to **examine** the address of the contact for an *ON* condition. If the N.O. contact has an input address, and if the real-world input device is closed, or *ON*, the processor sets the appropriate bit in the input register to 1, or *ON*. As the EXAMINE ON instruction is looking for an *ON* condition, a bit set to 1, or *ON*, is a true condition, and a logic path exists through the contacts. If the real-world input had been open, or *OFF*, the processor would have cleared the appropriate bit to 0, or *OFF*, and the contact would be false as far as the logic of the ladder diagram was concerned and would not allow a logic path.

EXAMINE OFF
When the N.C. symbol is programmed in a ladder diagram, the processor views it as a request to **examine** the address of the contact for an *OFF* condition. Any address that is actually *OFF* becomes logically true and power can flow. If a N.C. contact has an input address and the real-world input device is open, or *OFF*, the processor sets the bit to 0, or *OFF*. As the EXAMINE OFF instruction is looking for an *OFF* condition, a bit set to 0, or *OFF*, is a true condition and there would be logic continuity, so power can flow through the contact. If the input device had been closed, or *ON*, the bit would be set to 1, or *ON*. The EXAMINE OFF instruction can only be logically true when an *OFF* condition exists. Any bit set to 1 is viewed as an *ON* condition which makes an EXAMINE OFF (N.C.) contact false and no power can flow.

To further reinforce the EXAMINE ON and EXAMINE OFF instruction concepts, a look at a standard *STOP/START* station may be helpful. Figure 8–7a shows a standard Stop/Start circuit with overload contacts using a standard ladder diagram. Figure 8–7b shows the equivalent circuit program for a PLC.

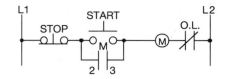

Figure 8–7a Standard *STOP/START* Ladder

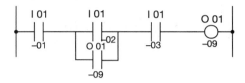

Figure 8–7b Equivalent *STOP/START* Circuit Programmed with a PLC

NOTE: Addresses shown are Square D format. An "I" preceding a word and bit number indicates an input whereas an "O" preceding a word and bit address indicates an output. Since the output must be the last item programmed on a Rung, the overload contacts (O.L./I01-03) are programmed ahead of the motor.

Once the input devices are wired to the input module(s) as shown in Figure 8–7c, and the PLC system is "powered up", or turned *ON*, the processor scans the inputs and sets the corresponding bits to 1 or 0 depending on the status of the real-world input devices. If an input is open, the corresponding bit is set to 0, or *OFF*, whereas any bit that represents a closed device will be set to 1, or *ON*.

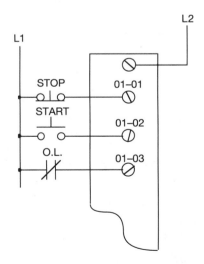

Figure 8–7c Input Devices Connected to an Input Module

In Figure 8–7c, the *STOP* button (I01-01) and the O.L. contact (I01-03) are programmed as normally opened contacts (EXAMINE ON). Because the *STOP* button and overload contacts are actually closed, bits 01 and 03 of word 01 are set to 1, or *ON*. Figure 8–8a shows the bit status of word 01 with the processor in the *RUN* mode.

16	15	14	13	12	11	10	9	8	7	6	5	4	3	2	1
0	0	0	0	0	0	0	0	0	0	0	0	0	1	0	1

Figure 8–8a Bit Status of Word 01 with Processor in *RUN* Mode

With the *STOP* button and the O.L. contacts (bits 01 and 03), set to 1, or *ON*, we need only press the *START* button to complete the circuit. When the *START* button (I01-02) is depressed, bit 02 is set to 1 (*ON*) during the next processor scan (shown in Figure 8–11b), and the circuit logic to the output is complete.

16	15	14	13	12	11	10	9	8	7	6	5	4	3	2	1
0	0	0	0	0	0	0	0	0	0	0	0	0	1	1	1

Figure 8–8b Bit Status of Word 01 while Start Button is being Depressed

With the circuit logic now complete (true), the processor sets bit 09 of word 01 to 1, or *ON* (Figure 8–8c), and motor 001-09 is *energized* and held energized by holding contacts 001-09. The holding contacts O01-09 have the same address as the motor (O01-09) because they both reference the same bit (09) of word 01 in the I/O register. The holding contacts do not actually exist as hard-wired contacts, but are used to maintain the logic path for the circuit. The EXAMINE ON instruction with the address O01-09 is the equivalent of holding contacts, and when the motor is energized (turned *ON*), bit 09 of word 01 set to 1, or *ON*, and an alternate logic path is now complete.

16	15	14	13	12	11	10	9	8	7	6	5	4	3	2	1
0	0	0	0	0	0	0	1	0	0	0	0	0	1	0	1

Figure 8–8c Bit Status of Word 01 after Output 001–09 is Energized and Start Button is Released

When the *STOP* button (input I01-01) is depressed, the processor clears bit 01 to 0, and the circuit logic is broken, or goes FALSE. Bit 09 is cleared to 0, and the real-world output device connected to terminal 01-09 drops out (turns *OFF*). Word 01 in the I/O register now appears as it is shown in Figure 8–8d, with only bit 03, the overload contact, set to 1.

16	15	14	13	12	11	10	9	8	7	6	5	4	3	2	1
0	0	0	0	0	0	0	0	0	0	0	0	0	1	0	0

Figure 8–8d Bit Status of Word 01 with Stop Button Depressed

When the *STOP* button is released or closed again, bit 01 is again set to 1, or *ON* (Figure 8–8e). The output (001-09) is not energized, however, as the *START* button (bit 02) is 0 (*OFF*), and the holding contacts (bit 09) are cleared to 0 (*OFF*).

16	15	14	13	12	11	10	9	8	7	6	5	4	3	2	1
0	0	0	0	0	0	0	0	0	0	0	0	0	1	0	1

Figure 8–8e Bit Status of Word 01 with *STOP* Button Released

Another way to look at the normally open and normally closed symbols used for programming the PLC is **relay analogy**.

For the sake of discussion, imagine that each input device is connected to an invisible control relay inside the input module, and that each control relay has one normally open and one normally closed contact as shown in Figure 8–9.

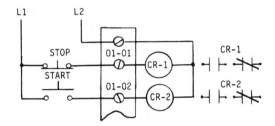

Figure 8–9 Imaginary Control Relays Wired to an Input Module

Connect one lamp to the normally open contacts, and another lamp to the normally closed contacts of CR-1, as shown in Figure 8–10.

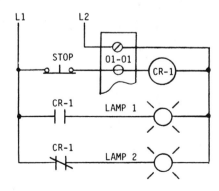

Figure 8–10 Lamps Wired to N.O. and N. C. CR-1 Contacts

When power is applied to L1 and L2, CR-1 energizes through the normally closed contacts of the *STOP* button. With CR-1 energized, the normally open contacts of CR-1 close, and lamp 1 lights as indicated in Figure 8–11. The normally closed contacts of CR-1 are now open, so lamp 2 *cannot* light.

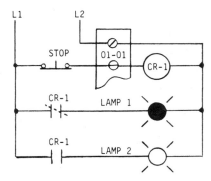

Figure 8–11 Lamp 1 Lights with Power Applied to L1 and L2

Wired in this manner, normally open CR-1 contacts controlled by a normally closed push button will close or *conduct* when power is applied to the circuit. Likewise, normally open contacts *programmed* to represent a normally closed push button will conduct when power is applied to the PLC.

A normally open *START* button connected to an input terminal and an invisible or imaginary control relay (shown in Figure 8–12) would not light Lamp 1 until the *START* button was depressed, and would only stay lit as long as the button was held down. Lamp 2 would light as soon as power was applied, but would go out when the *START* button was depressed and CR-2 energized (Figure 8–13).

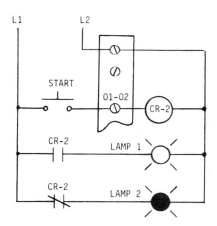

Figure 8–12 Power Applied—
START Button Not Depressed

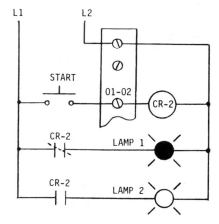

Figure 8–13 Power Applied—
START Button Depressed

The rules for contacts that represent real-world input devices are shown in Figures 8–14a and 8–14b.

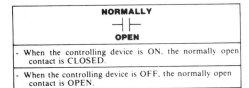

Figure 8–14a
(Courtesy of Square D Company)

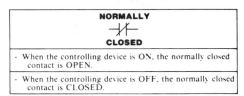

Figure 8–14b
(Courtesy of Square D Company)

There are no invisible control relays in the input modules, and there are also no symbols on the programming device for *STOP* buttons, *START* buttons, limit switches, and the like. As long as relay contact symbols must be used in place of regular input symbols, the relay analogy is an easy way to explain why normally open contacts are programmed to represent normally closed input devices.

It does not matter which approach you use to understand the logic behind the way that PLCs are programmed, as long as you DO understand it. The author believes that the EXAMINE ON and EXAMINE OFF approach is the easiest and clearest way to look at programming. The conversion of ladder diagram to PLC program will be quite simple once you clearly understand the concept of EXAMINE ON and EXAMINE OFF.

Clarifying EXAMINE ON and EXAMINE OFF

From the previous discussion and examples, it appears that all input devices, whether they are normally open or normally closed, are programmed as N.O. contacts to achieve the desired results in the ladder diagram. For many circuit applications this is true, and a big advantage of this programming technique is the ability to turn single-pole input devices into double-, three-, or four-pole devices in the circuit. Figure 8–15 shows a double-pole pressure switch (PS-1) controlling two outputs: motor 1 and motor 2.

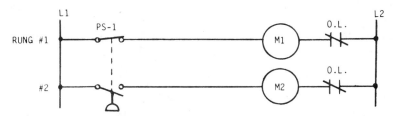

Figure 8–15 Double Circuit Pressure Switch

When power is applied to the circuit, motor 1 will start through the N.C. contacts of PS-1 in Rung 1. Motor 2 cannot start, however, due to the N.O. contacts of PS-1 in Rung 2. When PS-1 is actuated, the N.C. contacts in Rung 1 will open and motor 1 will go *OFF,* while the N.O. contacts in Rung 2 will close, turning motor 2 *ON.*

By programming the same circuit on a PLC, the necessity of buying a double circuit pressure switch is eliminated. One N.O. and 1 N.C. contact having the same address is used (see Figure 8–16a). The address is actually the address of a discrete N.C. single circuit pressure switch (illustrated in Figure 8–16b).

Figure 8–16a Double Circuit Pressure Switch
Circuit Programmed for a Typical PLC

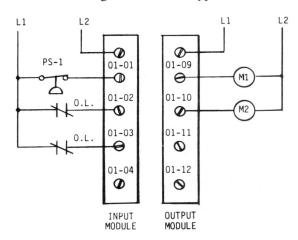

Figure 8–16b Actual Wiring of a Single-Pole
Pressure Switch

When the processor is placed in the *RUN* mode, it examines all N.O. contacts for an *ON* condition, or EXAMINE ON. As PS-1 and both overload contacts are closed (*ON*), bits 01, 02, and 03 are set to 1, making those portions of the ladder diagram true. When the processor EXAMINES *OFF* the N.C. contact (I01-01), it sees that bit 01 is set to 1 so that this part of the ladder diagram is false. Motor 1, address 001-09, is *ON* and motor 2, address 001-10, is *OFF.*

When the pressure switch is actuated, the processor continues scanning the user program and examines all N.O. contacts for *ON,* and all programmed N.C. contacts for *OFF.* Since the pressure switch (PS-1) is now actuated, the N.C. contacts are open, and the N.O. contacts (I01-01) go false, turning *OFF* motor 1 (001-09), whereas the N.C. contacts (I01-01) go true, and turn motor 2 (001-10) *ON*.

Even though only double, three, and four poles were mentioned, there is no limit (except user memory size) to the number of times an input device can be addressed and used in a programmed circuit. This programming technique allows for six-pole, seven-pole, eight-pole, and so on, devices to be programmed using only a single-pole discrete device.

There are many more applications and circuits that normally require two- or three-pole devices that now only require single-pole devices when programmed for a PLC. Examples are double-pole limit switches for forward and reversing circuits, and double-pole pressure switches for duplex controllers.

When programming contacts (N.O. or N.C.) which are controlled by outputs, the familiar standard relay logic is used. Figure 8–17a shows a standard *STOP/START* station with pilot lights. Lamp 1 (green) indicates power is available, and lamp 2 (red) indicates the circuit is activated. Figure 8–17b shows how the circuit is programmed.

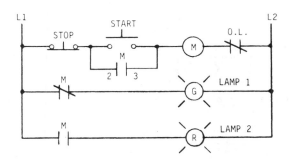

Figure 8–17a Ladder Diagram for *STOP/START* Station with Indicator Lamps

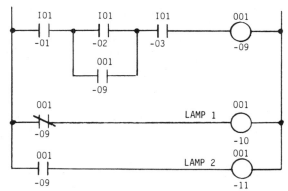

Figure 8–17b Programmed Circuit for *STOP/START* Station with Indicator Lamps

When the processor is placed in the *RUN* mode, lamp 1 lights due to the N.C. contacts 001-09. When the *START* button is depressed, output 001-9 energizes, N.C. contacts 001-09 open (go false), and lamp 1 (001-10) goes out. N.O. contacts 001-09 close (go true), lamp 2 (001-11) turns *ON*, and the holding contacts (001-09) go true to complete the circuit logic.

Notice that the output, holding contacts, N.C. contact, and N.O. contact for the lamps all have the same address (001-09). The address refers to bit 09 of word 01, and this same bit (09) is referenced four times in the ladder diagram. The EXAMINE OFF, or N.C. contact, is logically true when bit 09 is cleared to 0 (*OFF*), whereas the EXAMINE ON, or N.O. contacts, are not logically true until bit 09 is set to 1 (*ON*).

Figure 8–18a shows the bit status for the circuit with power applied; Figure 8–18b, with the *START* button depressed; Figure 8-18c, with the *START* button released; Figure 8-18d, with the *STOP* button depressed; and Figure 8-18e, after *STOP* button is released.

16	15	14	13	12	11	10	9	8	7	6	5	4	3	2	1
0	0	0	0	0	0	1	0	0	0	0	0	0	1	0	1

Figure 8–18a Bit Status of Word 01 with Power Applied

16	15	14	13	12	11	10	9	8	7	6	5	4	3	2	1
0	0	0	0	0	1	0	1	0	0	0	0	0	1	1	1

Figure 8–18b Bit Status of Word 01 with *START* Button Depressed

16	15	14	13	12	11	10	9	8	7	6	5	4	3	2	1
0	0	0	0	0	1	0	1	0	0	0	0	1	0	1

Figure 8–18c Bit Status of Word 01 with *START* Button Released

16	15	14	13	12	11	10	9	8	7	6	5	4	3	2	1
0	0	0	0	0	0	1	0	0	0	0	0	0	1	0	0

Figure 8–18d Bit Status of Word 01 with *STOP* Button Depressed

16	15	14	13	12	11	10	9	8	7	6	5	4	3	2	1
0	0	0	0	0	0	1	0	0	0	0	0	0	1	0	1

Figure 8–18e Bit Status of Word 01 with *STOP* Button Released

Chapter Summary

The relay logic used for programming the PLC at first seems to be in conflict with standard ladder logic. But once the concepts of EXAMINE ON and EXAMINE OFF are understood, the logic process is easy to understand. An EXAMINE ON instruction looks for an *ON* condition, and will be logically true when an *ON* condition (a 1) is found. The EXAMINE OFF instruction will be logically true when an *OFF* condition (a 0) is found. Looking at the actual status of the bits within individual I/O words is another way to help understand how the processor logic works. Others find that the relay analogy approach to understanding the normally open (EXAMINE ON) and normally closed (EXAMINE OFF) symbols used for programming is better. Another approach is to accept the fact that an N.C. *STOP* button, or a similar closed input device must be programmed using an N.O. contact symbol, and just have the philosophy that logic is relative to application and don't worry about it.

REVIEW QUESTIONS_____

1. Briefly describe the action of the EXAMINE ON instruction.
2. When a normally open (N.O.) limit switch is wired to an input module, and programmed using a N.O. contact symbol (EXAMINE ON), the instruction will be true when (check all correct answers):
   a. power is applied and the key switch is in the *RUN* position.
   b. the limit switch is closed.
   c. as long as the limit switch is open
   d. never
3. If the normally open limit switch in Question 1 is programmed using a N.C. contact symbol (EXAMINE OFF), the instruction will be true when (check all correct answers):
   a. power is applied and the key switch is in the *RUN* position.
   b. the limit switch is closed.
   c. as long as the limit switch is open
   d. never
4. Briefly describe the action of the EXAMINE OFF instruction.
5. Indicate the logic, (T [True] or F [False]) for the following contacts:

CONDITION OF INPUT DEVICE	PROGRAM INSTRUCTION	LOGIC TRUE-FALSE
a.		T F
b.		T F
c.		T F
d.		T F

Chapter 9

Programming a PLC

Objectives

After completing this chapter you should have the knowledge to
  • Explain the term *on line programming*.
  • Describe basic programming techniques.
  • Describe the FORCE ON and FORCE OFF features, and the hazards that could be associated with both.

The electrical symbol keys on dedicated programmers that represent N.O. contacts, N.C. contacts, branch circuit start and end, coils/outputs, timers, counters, and so forth, are called **OP codes** (operating codes) to tell the processor what to do.

NOTE: The terms "branch circuit start" and "branch end" refer to the start and end of circuit junctions where contacts are connected in parallel.

With a 16-bit word of memory, four bits are usually used for an OP code to tell the processor *what* to do, and the remaining 12 bits are the address, and tell the processor *where* to do it. When the key for a normally open contact symbol (EXAMINE ON) is pushed and then followed by an address, one word of user memory is used. Four bits tell the processor to treat this as an N.O. contact, or a request to EXAMINE ON. The next twelve bits hold the address or location of the input, output, or internal location with which the contact is associated.

For most PLC systems, each N.O. and N.C. contact, branch start and end, and coil (output) instruction requires one word of user memory, while timers and counters require from two to five, depending on the PLC. In reality, when a contact is entered and addressed, one full word of user memory is used. Many programmers display the total number of memory words used on the Video Display Terminal (VDT). The total also includes any words of memory, if any, that the processor used for internal purposes.

Since the user program requires one full word for each contact or coil, and two or more words for each timer and/or counter programmed, it is not unusual for the user memory to use more of the total memory words than the storage memory, since the storage memory only uses 1 bit of a 16-bit word to store the status of the various input and output devices.

For processors with 8-bit words, contacts will normally use two words, while timers or counters and coils will use four or more words each.

Programmers that can create, modify, monitor, and load programs into user memory can also make changes to the program while the processor and driven equipment is running. This feature is often referred to as **on line programming**. Changing the program while the processor is running *must only be done* by persons with a complete understanding of not only the circuit operation, but also the process or driven equipment as well. To prevent unauthorized on line programming, a key switch is provided either on the programming device or on the processor. With the switch placed in the *RUN* position and the key removed, the programmer cannot be used for on line programming. The key switch can also restrict the programmer to a "monitor only" mode or to **off line** and monitor only mode. Off line programming, which means that the program is being developed off line (without being connected to the process or driven equipment) is the most common method of programming, and, of course, the safest. Since few programs are ever created without mistakes, it is best to always create the program off line. Once the program is complete, it should be tested while still in the off line mode. After testing and verifying the program (circuit) in off line, the PLC can be put in the on line mode for final testing and operation.

The function or functions of the programmer that can be locked out with the key switch are fairly standard, but may vary from PLC to PLC.

Some processors also use **passwords** to limit the access to the processor program. Passwords act like a key switch. If the wrong password is entered when requested, the user is denied further access to the program.

Sample Programs

Figure 9–1 shows the keyboard for the Square D Company class 8010 SY/MAX dedicated desktop programmer. The keyboard consists of standard alphanumeric keys in typewriter format. All the keys except the *CTRL* and *SHIFT* keys generate standard ASCII code characters. Cursor movement control and numeric keys are located to the right of the keyboard, and a row of multiple function **soft keys** are located across the top. Soft keys are keys that perform different functions, depending on the programmer mode.

Figure 9–1 Keyboard of a Square D Company SY/MAX Programmer

When the programmer is first turned on, it enters the initial mode, and the VDT display is shown (Figure 9–2). The information at the bottom of the screen indicates the function of the 10 soft keys. In the *initial* mode, only the first six keys are functional. By pressing the first soft key, *STATUS*, the programmer goes to the status display (Figure 9–3). From the *status* mode, the first key can be pressed again to enter the *ladder* mode. From the *ladder* mode, either *SEARCH*, *DELETE*, or *PROGRAM* modes may be selected.

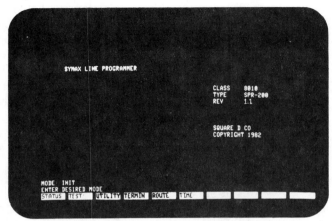

Figure 9–2 Initial Mode VDT Display
(Courtesy of Square D Company)

STATUS

PRESS ☐

The result will be:

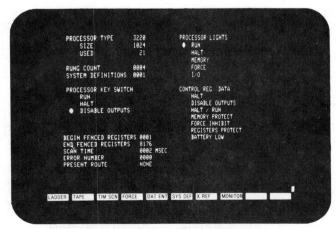

Figure 9–3 Status Mode VDT Display
(Courtesy of Square D Company)

To actually program a circuit using the programmer, the *initial* mode soft key is pressed, followed by the *status* soft key, then the *ladder* soft key, and then the *program* soft key. The VDT now displays the *program* mode as shown in Figure 9–4.

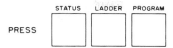

PRESS

The result will be:

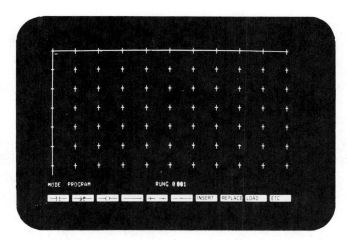

Figure 9–4 Program Mode Display
(Courtesy of Square D Company)

The soft keys now represent N.O. contacts, N.C. contacts, coil (output), space (no contact), open, and branch circuit *START/CLOSE*. The last four keys are: *INSERT;* for inserting a rung into the ladder diagram; *REPLACE* for replacing a rung in the ladder diagram; *LOAD* for loading the programmer rung(s) into memory; and *ETC*. When the *ETC* key is pressed, it changes the soft keys for programming timers, counters, and other special programming features.

By using the basic relay-type instruction keys (Figure 9–5a), the circuit can be programmed as shown in Figure 9–5b.

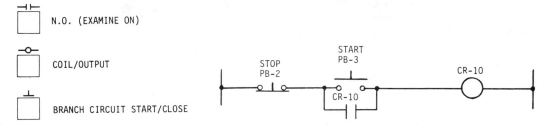

Figure 9–5a Relay-Type Instruction Keys   **Figure 9–5b** Standard Stop/Start Circuit

The **cursor**, which is controlled by the cursor movement keys at the top of the numeric key section of the keyboard, is used to indicate position on a rung. Initially, the cursor is at the extreme top left-hand portion of the screen. This is Rung 1, the first horizontal contact position.

The first step in programming the circuit in Figure 9–5b is to power up the programmer and press the "soft key" sequence: *STATUS*, *LADDER*, and *PROGRAM* (Figure 9–6).

Figure 9–6 Proper Keying Sequence to Enter Program Mode

The VDT now appears as shown in Figure 9–7. Notice the cursor is at the top left of the circuit matrix.

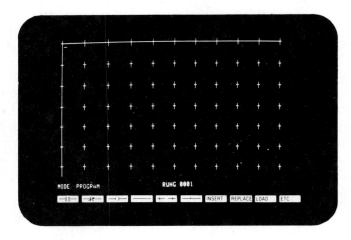

Figure 9–7 VDT Display for Program Mode
(Courtesy of Square D Company)

By pressing the key sequence shown in Figures 9–8a, 9–8b, and 9–8c, the circuit in Figure 9–5b can be programmed.

PRESS

The result will be:

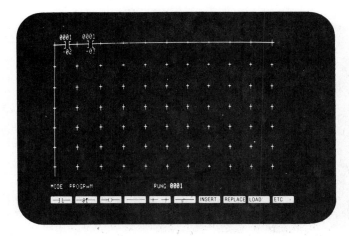

Figure 9–8a Key Strokes for Entering and Addressing the *STOP* and *START* Buttons (PB-2 and 3) and the Resulting VDT Display *(Courtesy of Square D Company)*

PRESS

The result will be:

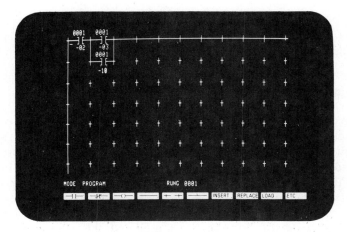

Figure 9–8b Key Strokes for Entering and Addressing the Parallel Holding Contacts and the Resulting VCT Display *(Courtesy of Square D Company)*

PRESS

The result will be:

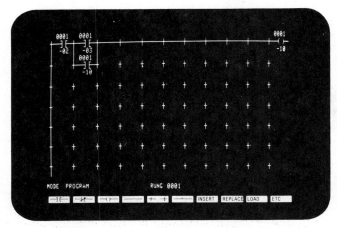

Figure 9–8c Key Strokes for Entering and Addressing the Output and the
Resulting VDT Display *(Courtesy of Square D Company)*

NOTE: The "soft keys" are indicated by a symbol or word *above* the keys whereas dedicated, or "hard keys," have the character *on* the keys.

The final step is to press the *LOAD* soft key twice (Figure 9–9). The first time the key is pressed, the circuit is formatted; the second time the key is pressed, the formatted circuit is entered into user memory.

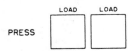

PRESS

Figure 9–9 Key Sequence for Loading Rung into Memory

It is not the purpose of this text to go into great detail on programming large or complex circuits, but rather to illustrate some simple programming methods so that the electrician or technician can get a *feel* for how relatively simple programming is. Each manufacturer uses a somewhat different technique, and the only way you can really learn to program a given PLC is to spend time on *that* PLC. While the techniques for programming differ, basic concepts are the same, and it is these concepts that this book covers.

Figure 9–10a shows the sealed touchpad keyboard used with the Allen-Bradley industrial terminal, and Figure 9–10b shows the overlays that fit over the sealed touchpads to identify the keys. Figure 9–10c illustrates how an overlay is installed on the keyboard.

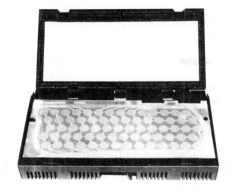

Figure 9–10a Sealed Touchpad Keyboard
(Courtesy of Allen-Bradley)

Figure 9–10b Keyboard Overlays
(Courtesy of Allen-Bradley)

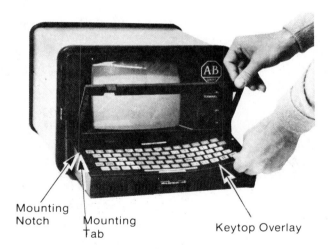

Mounting
Notch   Mounting
Tab   Keytop Overlay

Figure 9–10c Installing an Overlay on the Keyboard
(Courtesy of Allen-Bradley)

The overlay shown in Figure 9–11 is for programming and editing circuits using a T3 industrial terminal. The key groupings include numerics with force *ON* and *OFF* instructions, relay-type instructions, timer and counter instructions, data manipulation, arithmetic instructions, editing instructions, control instructions, and so forth. Additional overlays are available with alphanumeric keys (standard typewriter format) and alphanumeric and/or graphic modes.

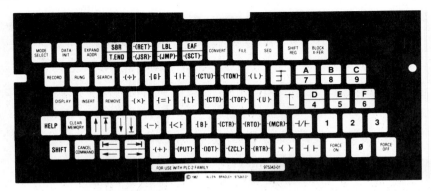

Figure 9–11 Programming and Editing Overlay
(Courtesy of Allen-Bradley)

When the programmer is connected to a Mini-PLC 2/15 and turned on, the mode select display appears as shown in Figure 9–12. For programming a circuit, the *processor* mode must be selected. This is done by entering the number 11 on the keyboard. The CRT now displays an empty screen with the words *START* and *END* (Figure 9–13). The *START* remains at the front of any circuit programmed, and the *END*, which indicates the end of the circuit, automatically moves down as contacts and/or rungs are added.

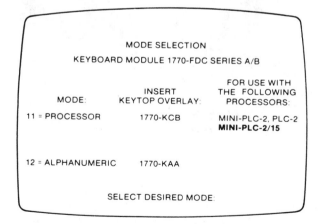

Figure 9–12 Initial VDT Display
(Courtesy of Allen-Bradley)

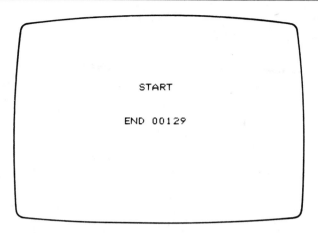

Figure 9–13 Processor Mode VDT Display

Notice that following the *END* statement is the number 00129. This number represents the total memory words used. One word is for the *END* statement, and 128 words are for the data table. Memory structure is explained in detail in Chapter 5. If the I/O table was configured for 256 words, the number shown would be 257 (one word for *END* and 256 words for the data table). As a program is entered, the number of additional memory words used is added to the total (Figures 9–16a, 9–16b, 9–16c, and 9–16d).

The basic relay instructions are shown in Figure 9–14.

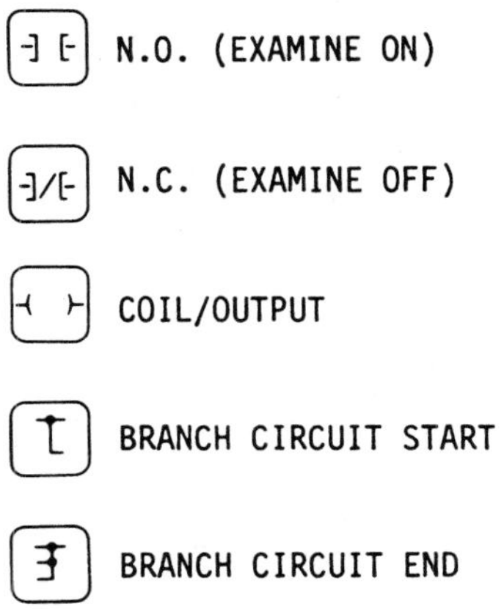

┤ ├   N.O. (EXAMINE ON)

┤/├   N.C. (EXAMINE OFF)

┥ ┝   COIL/OUTPUT

┌ T ┐   BRANCH CIRCUIT START

┌ ╀ ┐   BRANCH CIRCUIT END

Figure 9–14 Relay Instructions

To program the circuit shown in Figure 9–15 into user memory, the key sequences shown in Figures 9–16a, 9–16b, 9–16c, and 9–16d are used.

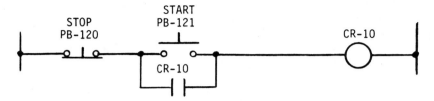

Figure 9–15 Standard *STOP/START* Circuit

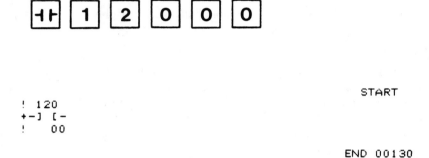

```
                                                          START

! 120
+-] [-
!   00
                                                          END  00130
```

Figure 9–16a Key Sequence for Entering and Addressing *STOP*
Button PB-120 and the Resulting VDT Display

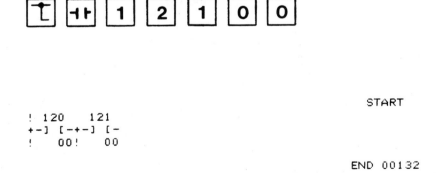

```
                                                          START

! 120    121
+-] [-+-] [-
!   00!   00
                                                          END  00132
```

Figure 9–16b Key Sequence for Entering and Addressing *START*
Button PB-121 and the Resulting VDT Display

```
LADDER DIAGRAM DUMP                                         START

! 120    121
+-] [-+-] [-+
!   00!   00!
!     ! 010 !
!     +-] [-+
!       00
                                                        END  00135
```

Figure 9–16c Key Sequence for Entering and Addressing Parallel
Holding Contacts and Resulting VDT Display

```
                                                           START
                                                                           010  !
! 120    121
+-] [-+-] [-+---------------------------------------------( )--+
!   00!   00!                                                           00  !
!     ! 010 !                                                               !
!     +-] [-+                                                               !
!       00                                                                  !

                              END  00136
```

Figure 9–16d Key Sequence for Entering and Addressing Output and
Resulting VDT Display

NOTE: When using an Allen-Bradley PLC-2 family, a 1 preceding an address indicates an input; an O preceding an address indicates an output. In Figure 9–16a the *STOP* button is given the address 12000 which is bit 00 of word 120 of the input image table, and the output address 01000 (Figure 9-16d) is bit 00 of word 010 of the output image table. (Refer to Figure 5–11 in Chapter 5.)

NOTE: The number following the *END* statement in Figure 9–16d indicates that 136 words of the total memory have now been used. Before programming was started, the number was 129 which included 128 words for the data table and one word for the *END* statement. Subtracting this total, 129, from total memory words, 136, gives us the total user memory words of seven (7) used for the program.

1 word for contact 12000
1 word for branch circuit start
1 word for contact 12100
1 word for branch circuit start
1 word for contact 01000
1 word for branch circuit end
1 word for coil 01000
---
7 total words of user memory.

Most PLCs are designed so that any input or output contact (N.O. or N.C.) or output coil can be FORCED ON or OFF. This feature (and troubleshooting aid) allows the operator to force *ON*, or make a contact or coil go *ON*, regardless of actual input device status or circuit logic. Similarly, contacts can be forced, or turned *OFF*, regardless of actual input device status or circuit logic.

> ***CAUTION:** The FORCE ON and FORCE OFF feature should *never* be used except by personnel who completely understand the circuit *and* the process machinery or driven equipment. An understanding of the potential effect that forcing a given contact or coil has on machine operation is essential if hazardous and/or destructive operation is to be avoided.

Programming with a Computer

Using a personal computer and software for programming the PLC requires knowledge of the computer, as well as the software. While additional skills are required, the benefits of the additional features offered by most software packages makes the time it takes to learn to operate the computer and to master the software time well spent.

To illustrate the differences in programming between a dedicated programmer and a personal computer with software, the Allen-Bradley 6200 Series Development and Documentation Software for the PLC-5 family of PLCs will be used.

The AB software uses the function keys on the computer in the same way that Square D used the soft keys. The software is designed so that various function keys throughout the programming process are used to call up various instructions when needed. These instructions include, but are not limited to, EXAMINE ON, EXAMINE OFF, output, and the like.

Once the software has been loaded, the opening screen appears as shown in Figure 9–17.

The PLC-5 version of the software is selected by pressing function key F5. Figure 9–18 shows the computer screen after the F5 function key has been pressed.

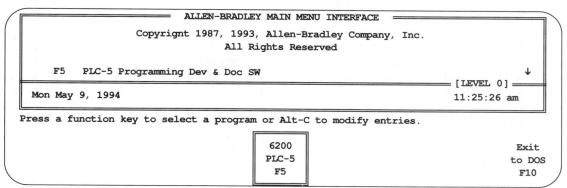

Figure 9–17 Opening Screen Using Allen-Bradley PLC Software
(Courtesy of Allen-Bradley)

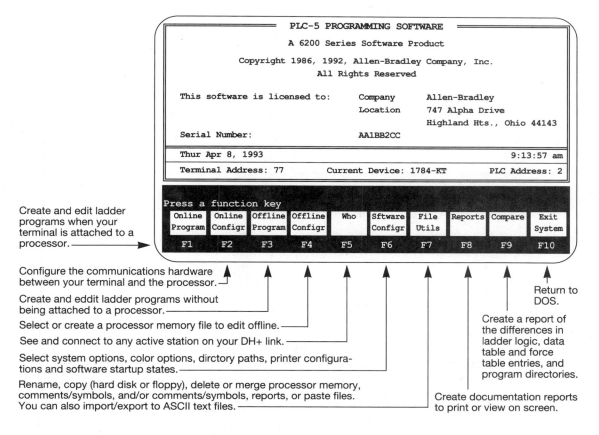

Figure 9–18 PLC-5 Programming Software Screen
(Courtesy of Allen-Bradley)

Once this screen is visible, the operator (programmer) is asked to select the desired mode from the **menu** at the bottom of the screen. The term menu is used to indicate the choices or functions that may be selected, much the same way we select food items from a restaurant menu. The choice for programming is offline and online. Offline programming is used when the computer running the software is not connected to the PLC processor, or when it is connected to the processor, but the program is developed (written) without connection to the actual equipment. Online programming allows the operator, or programmer, to make changes to the program while the program and equipment are actually running. On line programming and/or changes are normally restricted to experienced personnel only. For purposes of demonstration, we have selected off line configuration (F4), which is the first step necessary for off line programming. Figure 9–19 shows the resultant screen and the new menu at the bottom.

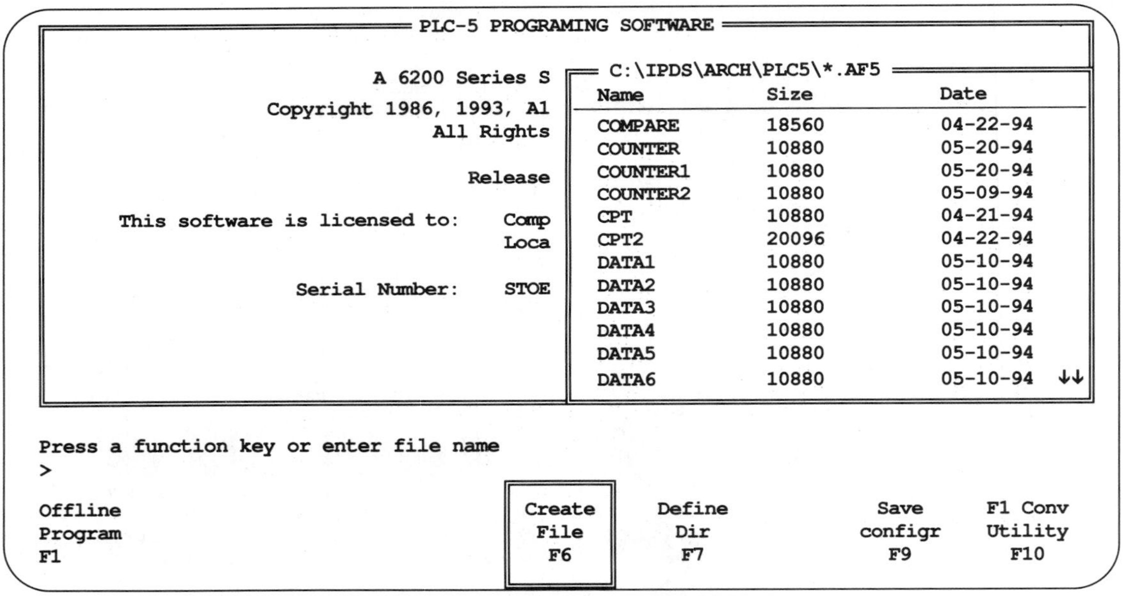

```
================= PLC-5 PROGRAMING SOFTWARE =================

                                    ┌═ C:\IPDS\ARCH\PLC5\*.AF5 ═══════════════
                A 6200 Series S     │ Name          Size          Date
        Copyright 1986, 1993, A1    │
                    All Rights      │ COMPARE       18560         04-22-94
                                    │ COUNTER       10880         05-20-94
                        Release     │ COUNTER1      10880         05-20-94
                                    │ COUNTER2      10880         05-09-94
     This software is licensed to:  Comp │ CPT      10880         04-21-94
                               Loca │ CPT2          20096         04-22-94
                                    │ DATA1         10880         05-10-94
             Serial Number:    STOE │ DATA2         10880         05-10-94
                                    │ DATA3         10880         05-10-94
                                    │ DATA4         10880         05-10-94
                                    │ DATA5         10880         05-10-94
                                    │ DATA6         10880         05-10-94  ↓↓

Press a function key or enter file name
>

Offline              Create      Define          Save       F1 Conv
Program              File        Dir             configr    Utility
F1                   F6          F7              F9         F10
```

Figure 9–19 Screen Display, Once Function Key F4 was Pressed

Using this software, an existing file can be edited (modified) or a new file can be created. The file must be created before we can actually create a program. To create a file, function key F6 is pressed. The software now asks the programmer to give the file a name. File names are limited to eight alphanumeric characters (letters and numbers), with no spaces. Usually, the file name relates in some way to the program that is to be written. For this programming example, the file is named "DEMO." The file is named by typing DEMO using the computer keyboard, and then pressing the *ENTER* key. Figure 9–20 shows the screen after the file has been named, and Figure 9–21 shows the screen after the *ENTER* key has been pressed.

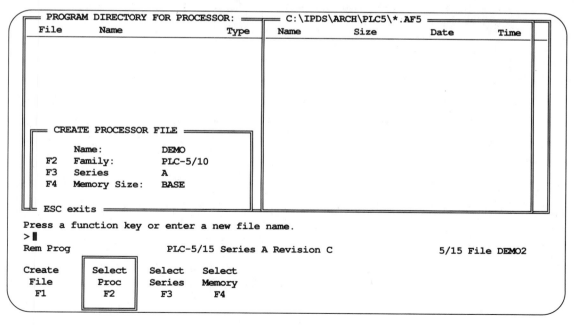

```
╒═══════════════ PLC-5 PROGRAMING SOFTWARE ══════════════════╕
│                                                             │
│              A 6200 Series S ╒══ C:\IPDS\ARCH\PLC5\*.AF5 ══╗ │
│          Copyright 1986, 1993, A1 ║ Name      Size      Date     ║ │
│                     All Rights ║                                 ║ │
│                               ║ COMPARE    18560     04-22-94  ║ │
│                               ║ COUNTER    10880     05-20-94  ║ │
│                     Release    ║ COUNTER1   10880     05-20-94  ║ │
│                               ║ COUNTER2   10880     05-09-94  ║ │
│  This software is licensed to:   Comp ║ CPT        10880     04-21-94  ║ │
│                              Loca ║ CPT2       20096     04-22-94  ║ │
│                               ║ DATA1      10880     05-10-94  ║ │
│          Serial Number:     STOE ║ DATA2      10880     05-10-94  ║ │
│                               ║ DATA3      10880     05-10-94  ║ │
│                               ║ DATA4      10880     05-10-94  ║ │
│                               ║ DATA5      10880     05-10-94  ║ │
│                               ║ DATA6      10880     05-10-94 ↓↓║ │
│                               ╚════════════════════════════════╝ │
│                                                             │
│ Enter File Name (No EXT)>                                   │
│ > Demo                                                      │
╘═════════════════════════════════════════════════════════════╛
```

Figure 9–20 Screen Once File has Been Named

```
╒═ PROGRAM DIRECTORY FOR PROCESSOR: ══╤═ C:\IPDS\ARCH\PLC5\*.AF5 ═╕
│ File     Name               Type │ Name     Size    Date    Time │
│                                  │                               │
│                                  │                               │
│                                  │                               │
│                                  │                               │
│                                  │                               │
│  ╒═ CREATE PROCESSOR FILE ═══╕    │                               │
│  │    Name:       DEMO        │   │                               │
│  │ F2 Family:     PLC-5/10    │   │                               │
│  │ F3 Series:     A           │   │                               │
│  │ F4 Memory Size: BASE       │   │                               │
│  ╘═ ESC exits ════════════════════╛                               │
│ Press a function key or enter a new file name.                    │
│ >█                                                                 │
│ Rem Prog           PLC-5/15 Series A Revision C      5/15 File DEMO2 │
│                                                                   │
│ Create    Select    Select   Select                               │
│  File      Proc     Series   Memory                               │
│   F1        F2        F3       F4                                  │
╘═══════════════════════════════════════════════════════════════════╛
```

Figure 9–21 Screen After Enter Key is Pressed

Once the file has been named, it is necessary to select the PLC-5 processor that will be used. The processor is selected by repeatedly pressing the F2 key until the correct processor is displayed in the lower left box (shown in Figure 9–21). For this example, the PLC-5/15 processor is selected. Figure 9–22 shows the screen with the PLC-5/15 Series A processor identified. The series type can be selected by pressing the *SELECT SERIES* key (F3).

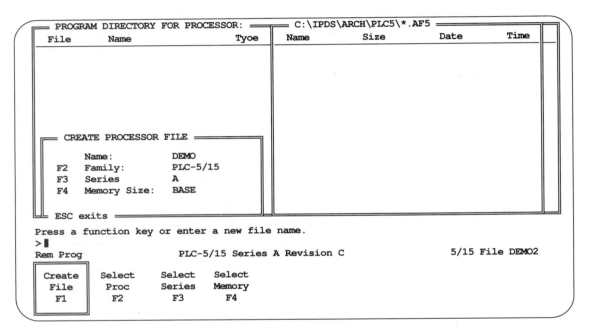

Figure 9–22 Screen with PLC-5/15 Selected

The next step is to press F1 (*CREATE FILE*). This creates the files that are needed to begin programming (Figure 9–23).

NOTE: Each screen shown in this programming sequence is after a key has been pressed, with the next key to be used highlighted. Example; the next step in the programming process is to press the F1 key (Create file). Figure 9–23 shows the resultant screen after F1 is pressed. The Monitor File key (F8) is highlighted and is the next key in the sequence that will need to be used. The next screen (Figure 9–24) shows how the screen looks after the F8 key has been pressed and highlights the next key to be used F10 (Edit).

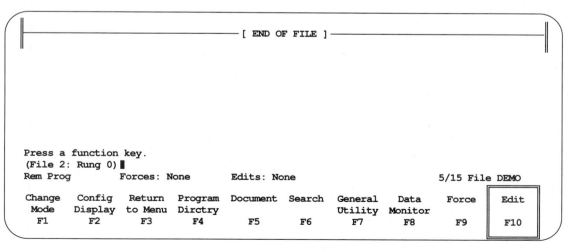

```
= PROGRAM DIRECTORY FOR PROCESSOR: DEMO2 ========================== [ OFFLINE ] =

   File        Name         Type         Size (words)

    0                       system            8
    1                       undefined         4
    2                       ladder            5
```

```
Press a function key or enter a new file name.
> █
Rem Prog                     PLC-5/15 Series A Revision C                  5/15 File DEMO

   Proc      Save/     Return    Change              Memory   General  Monitor
  Functns    Merge    to Menu     File                Map     Utility   File
    F1        F2        F3         F4                  F6        F7       F8
```

Figure 9–23 Creating Files for Program

Next, the F8 (*MONITOR FILE*) key is pressed, and produces the screen shown in Figure 9–24.

```
 ======================================== [ END OF FILE ] ========================================
```

```
Press a function key.
(File 2: Rung 0) █
Rem Prog          Forces: None        Edits: None                    5/15 File DEMO

 Change   Config   Return   Program  Document  Search  General    Data    Force    Edit
  Mode    Display  to Menu  Dirctry                    Utility  Monitor
   F1       F2       F3       F4        F5      F6        F7       F8       F9      F10
```

Figure 9–24 Screen Ready to Monitor File

We can now start to program by pressing the F10 (*EDIT*) key. Figure 9–25 shows the resultant screen.

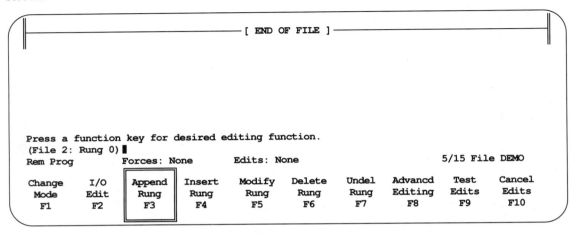

Figure 9–25 Screen Ready to Program

The F3 key is pressed twice (APPEND RUNG/APPEND INSTRUCTION) to start the actual programming process. Figure 9–26 shows the screen. (Note that the cursor is located in the upper left-hand corner of the screen.)

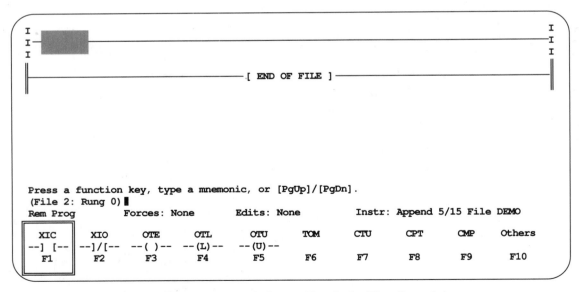

Figure 9–26 Screen with Cursor Ready for First Instruction

Before any actual programming takes place, a review of some of the Allen-Bradley PLC-5 instructions will be helpful. Allen-Bradley refers to instructions as "Instruction Sets." Some of the most common are listed below with their mnemonic names.

INSTRUCTION SETS

INSTRUCTION SET	MNEMONIC NAME
Examine ON	XIC
Examine OFF	XIO
Output Energized	OTE
Output Latched	OTL
Output Unlatch	OTU
Timer ON Delay	TON
Timer OFF Delay	TOF
Retentive Timer ON Delay	RTO
Reset	RES

PLC-5 Addressing Scheme

To program a motor *STOP/START* circuit (without overloads), the following key sequences are necessary:

- Select XIC (normally open EXAMINE ON contacts) by pushing F1

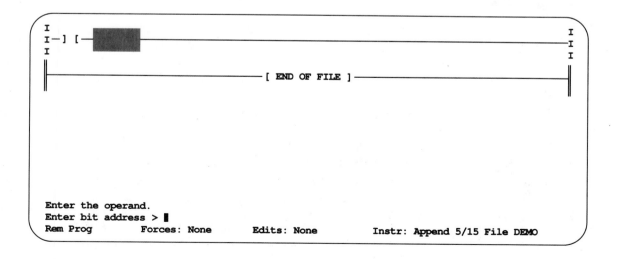

```
I                                                                    I
I—] [—                                                              —I
I                                                                    I
||——————————————————[ END OF FILE ]—————————————————————————||


 Enter the operand.
 Enter bit address > █
 Rem Prog        Forces: None        Edits: None        Instr: Append 5/15 File DEMO
```

- Enter address for *STOP* button using PLC-5 format   **i ; 001/04**
Input Word 001 Bit 04.

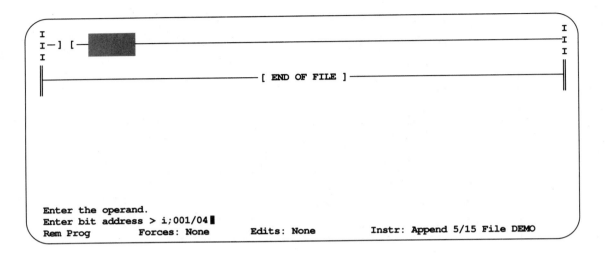

```
I                                                                    I
I —] [—  ███████                                                    —I
I                                                                    I
�photograph
║————————————————————————————[ END OF FILE ]————————————————————————║


  Enter the operand.
  Enter bit address > i;001/04█
  Rem Prog          Forces: None        Edits: None        Instr: Append 5/15 File DEMO
```

- Press enter key   *<ENTER>*

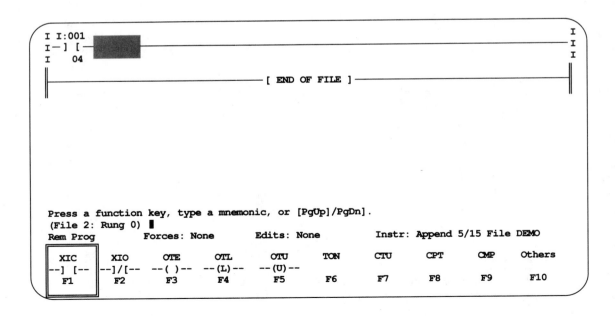

```
I I:001                                                              I
I —] [—  ███████                                                    —I
I   04                                                               I
║————————————————————————————[ END OF FILE ]————————————————————————║


  Press a function key, type a mnemonic, or [PgUp]/[PgDn].
  (File 2: Rung 0) █
  Rem Prog          Forces: None        Edits: None        Instr: Append 5/15 File DEMO

  ┌────────┐
  │  XIC   │   XIO      OTE      OTL      OTU      TON     CTU     CPT     CMP     Others
  │ --] [--│  --]/[--  --( )--  --(L)--  --(U)--
  │  F1    │   F2       F3       F4       F5       F6      F7      F8      F9      F10
  └────────┘
```

NOTE: To speed programming, the leading zeros can be omitted when entering an address; the zeros are automatically added by the software. The previous address i;001/04 could have been entered as i;1/4. The results on the screen would have been I:001/04. Although the letters are entered in lower case, the screen will display capital letters. A semicolon or a colon can be used after the i or o. The semicolon is normally used instead of the colon, because the colon requires that the *SHIFT* key be pressed.

- Press F1 again to select XIC (EXAMINE ON) contact
- Enter address for *START* button   **i;1/3**
  Input Word 001 Bit 03

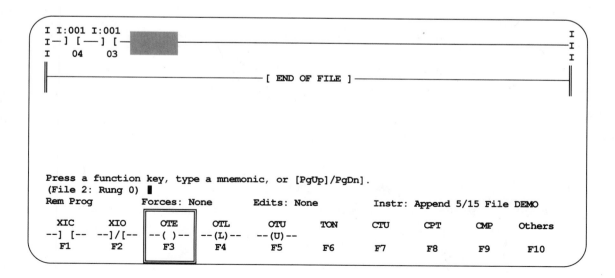

```
I I:001                                                               I
I—] [— ] [ ———————————————————————————————————————————————————————————I
I   04                                                                I
╟————————————————————————————— [ END OF FILE ] ———————————————————————╢




Enter the operand.
Enter bit address . i;1/3█
Rem Prog           Forces: None        Edits: None        Instr: Append 5/15 File DEMO
```

- Press enter key   *<ENTER>*

```
I I:001 I:001                                                         I
I—] [— ] [ —▓▓▓▓▓▓———————————————————————————————————————————————————I
I   04    03                                                          I
╟————————————————————————————— [ END OF FILE ] ———————————————————————╢




Press a function key, type a mnemonic, or [PgUp]/[PgDn].
(File 2: Rung 0) █
Rem Prog           Forces: None        Edits: None        Instr: Append 5/15 File DEMO

  XIC       XIO       OTE      OTL      OTU      TON     CTU     CPT     CMP     Others
--] [--   --]/[--   --( )--  --(L)--  --(U)--
  F1        F2        F3       F4       F5       F6      F7      F8      F9      F10
```

• Select *OTE* for the output using function key F3

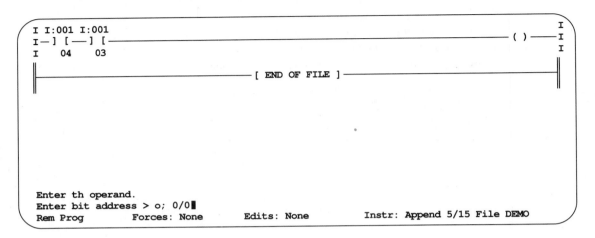

```
I I:001 I:001                                                              I
I—] [ —] [ —————————————————————————————————————————————— ( ) —— I
I   04    03                                                               I
                                                                         ‖
‖—————————————————————————————[ END OF FILE ]——————————————————————————‖
‖                                                                        ‖



Enter th operand.
Enter bit address > o; 0/0▮
Rem Prog            Forces: None        Edits: None      Instr: Append 5/15 File DEMO
```

• Enter address of output   **o ; 0 / 0   *RETURN***
  Output Word 000   Bit 00.
  The zeros will be added by the software to speed programming. An output word is
  indicated by the letter O, not a zero.

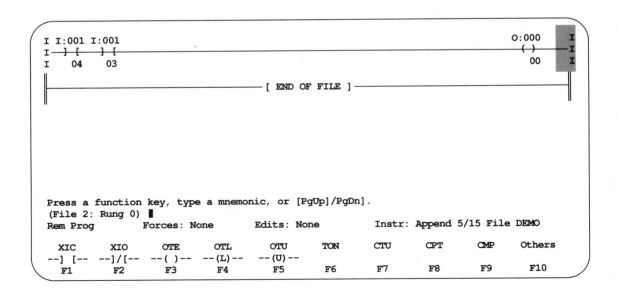

```
I I:001 I:001                                                   O:000    I
I—] [  —] [                                                      ( )   —I
I   04    03                                                      00     I
‖
‖—————————————————————————————[ END OF FILE ]——————————————————————————‖



Press a function key, type a mnemonic, or [PgUp]/[PgDn].
(File 2: Rung 0) ▮
Rem Prog            Forces: None        Edits: None      Instr: Append 5/15 File DEMO

   XIC       XIO       OTE       OTL       OTU      TON     CTU     CPT     CMP     Others
--] [--   --]/[--   --( )--   --(L)--   --(U)--
   F1        F2        F3        F4        F5       F6      F7      F8      F9      F10
```

- Use the cursor keys to move the cursor back to the instruction that is to be paralleled with the holding contact ← ←

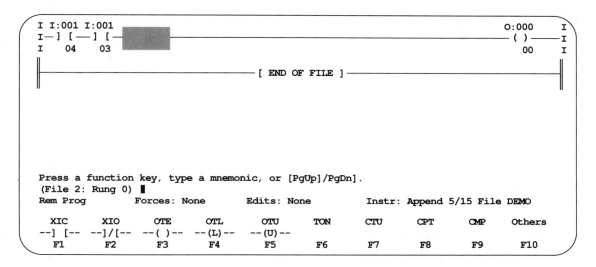

```
I I:001 I:001                                                    O:000   I
I—] [—] [—█████████———————————————————————————————————————————( )———I
I   04    03                                                      00    I
  ╠══════════════════════════[ END OF FILE ]═══════════════════════════╣



 Press a function key, type a mnemonic, or [PgUp]/PgDn].
 (File 2: Rung 0) █
 Rem Prog          Forces: None        Edits: None        Instr: Append 5/15 File DEMO

   XIC       XIO       OTE       OTL       OTU       TON     CTU     CPT     CMP    Others
 --] [--  --]/[--   -- ( )--  --(L)--   --(U)--
   F1        F2        F3        F4        F5        F6      F7      F8      F9     F10
```

- Press the escape key to change screens   *ESC*

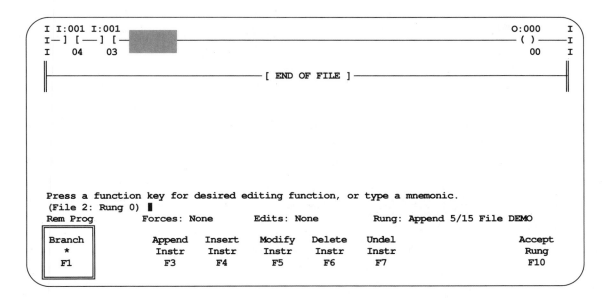

```
I I:001 I:001                                                    O:000   I
I—] [—] [—█████████———————————————————————————————————————————( )———I
I   04    03                                                      00    I
  ╠══════════════════════════[ END OF FILE ]═══════════════════════════╣



 Press a function key for desired editing function, or type a mnemonic.
 (File 2: Rung 0) █
 Rem Prog          Forces: None        Edits: None        Rung: Append 5/15 File DEMO

 ┌─────────┐
 │ Branch  │        Append   Insert   Modify   Delete   Undel              Accept
 │   *     │        Instr    Instr    Instr    Instr    Instr               Rung
 │  F1     │         F3       F4       F5       F6       F7                  F10
 └─────────┘
```

• Press F1 to select branch

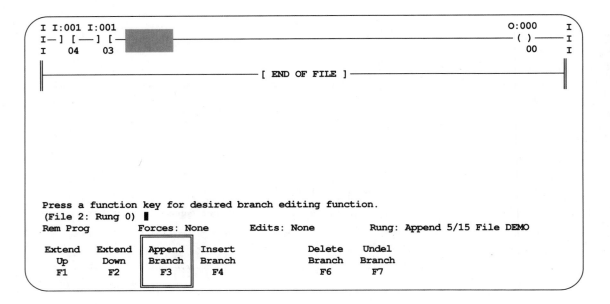

• To append (modify or change) the branch, the *APPEND BRANCH* key is pressed (F3)

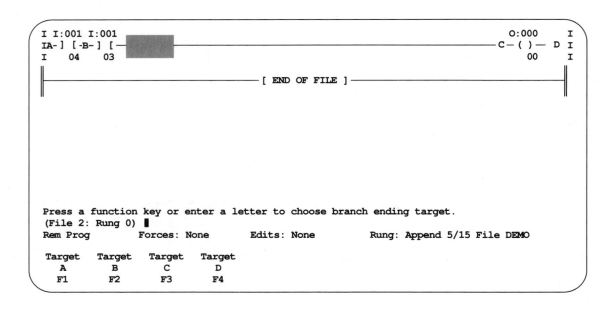

- Select target choice A, B, or C
  A is the *STOP* button **I:001/04**
  B is the *START* button **I:001/03**
  C is the *OUTPUT* **O:000/00**
- Target choice B is selected by pressing F2

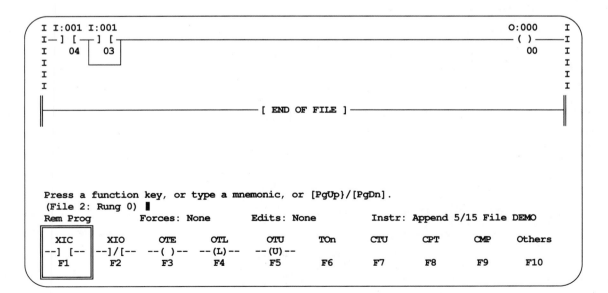

- *APPEND INSTRUCTION* is selected using function key F3

• Select XIC (EXAMINE ON) by pressing F1

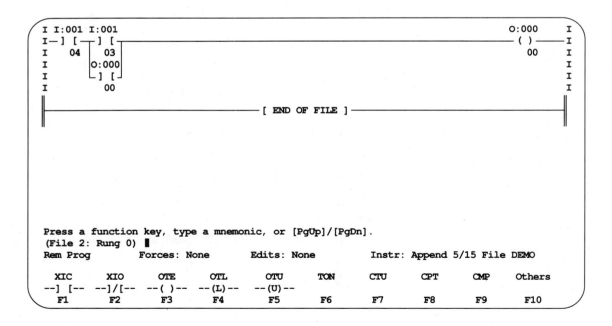

```
I I:001 I:001                                                    O:000    I
I─] [─┬─] [───────────────────────────────────────────────────── ( ) ───I
I   04 │  03                                                       00    I
╲I     └─] [─┘                                                           I
I                                                                       I
I                                                                       I
╟───────────────────────── [ END OF FILE ] ──────────────────────────╢

Enter the operand.
Enter bit address > o; 0:0█
Rem Prog          Forces: None        Edits: None         Instr: Append 5/15 File DEMO
```

• Enter address of holding contacts from the *OTE*   **o ; 0 / 0**   *<ENTER>*

```
I I:001 I:001                                                    O:000    I
I─] [─┬─] [───────────────────────────────────────────────────── ( ) ───I
I   04 │  03                                                       00    I
I     │O:000                                                             I
I     └─] [─┘                                                            I
I        00                                                             I
╟───────────────────────── [ END OF FILE ] ──────────────────────────╢

Press a function key, type a mnemonic, or [PgUp]/[PgDn].
(File 2: Rung 0) █
Rem Prog          Forces: None        Edits: None         Instr: Append 5/15 File DEMO

   XIC       XIO       OTE       OTL       OTU       TON      CTU      CPT      CMP      Others
--] [--   --]/[--   --( )--   --(L)--   --(U)--
   F1        F2        F3        F4        F5        F6       F7       F8       F9       F10
```

- To finish the program and accept it as it is displayed on the screen, the *ESCAPE* button is pressed to bring up a new screen   *ESC*

```
I I:001 I:001                                                          O:000    I
I─] [─┬─] [─────────────────────────────────────────────────( )────── I
I   04 │  03                                                            00    I
I      │O:000                                                                  I
I      └─] [─┘                                                                 I
I         00                                                                   I

╠═══════════════════════════════[ END OF FILE ]═══════════════════════════════╣


Press a function key for desired editing function, or type a mnemonic.
(File 2: Rung 0) ▐
Rem Prog          Forces: None          Edits: None          Rung: Append 5/15 File DEMO

Branch            Append   Insert   Modify   Delete   Undel                ┌─────────┐
  *               Instr    Instr    Instr    Instr    Instr                │ Accept  │
  F1              F3       F4       F5       F6       F7                    │  Rung   │
                                                                           │  F10    │
                                                                           └─────────┘
```

- Accept the rung by pressing F10
- Complete the program by pressing *ESCAPE* twice   *ESC   ESC*

The first reaction to the number of keystrokes required for programming using a computer and specialized software is, "WOW! What a lot of keystrokes!" While this method does take more keystrokes than are required when using a dedicated desktop programmer, the computer offers many advantages over the dedicated programmer, as discussed in Chapter 4. As with most new things learned, once the skill is mastered, the programming time is greatly reduced.

Another method of programming the same circuit that greatly reduces the keystrokes and reduces programming time uses the keyboard to enter the program rather than relying solely on function keys. Some function keys are still used, but not to the extent they were in the first example. This second method is sometimes referred to as the **shorthand method,** and requires that the electrician or technician knows the mnemonic names for each program instruction.

From the screen shown in Figure 9–24, the following keystrokes are used:
- *EDIT*   F10
- *APPEND RUNG*   F3

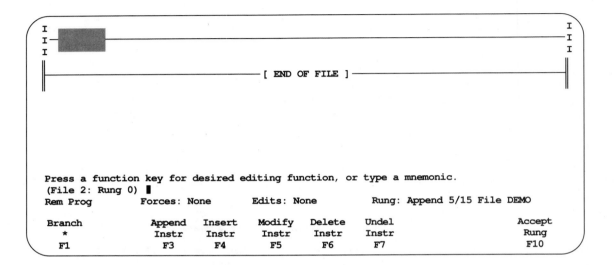

• Type instructions for first line of the rung   **xic i;1/4 xic i;1/3 ote o;0/0**

NOTE: All instructions and addresses are separated with a space.

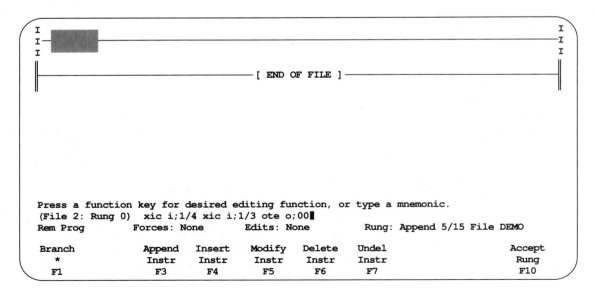

• Press *ENTER* key   *<ENTER>*

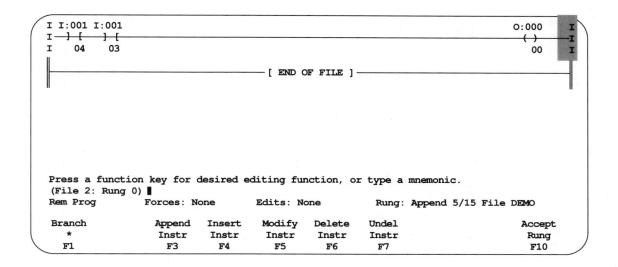

```
I I:001 I:001                                                    O:000    I
I—] [—] [——————————————————————————————————————————————————————( )——I
I   04    03                                                      00      I
 ‖—————————————————————————[ END OF FILE ]————————————————————————————‖

Press a function key for desired editing function, or type a mnemonic.
(File 2: Rung 0) ▮
Rem Prog          Forces: None        Edits: None        Rung: Append 5/15 File DEMO

Branch            Append    Insert    Modify    Delete    Undel                  Accept
  *               Instr     Instr     Instr     Instr     Instr                  Rung
  F1              F3        F4        F5        F6        F7                      F10
```

• Cursor back   ← ←

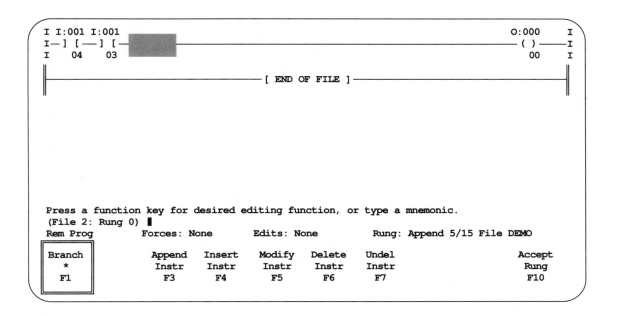

```
I I:001 I:001                                                    O:000    I
I—] [—] [                ——————————————————————————————————————( )——I
I   04    03                                                      00      I
 ‖—————————————————————————[ END OF FILE ]————————————————————————————‖

Press a function key for desired editing function, or type a mnemonic.
(File 2: Rung 0) ▮
Rem Prog          Forces: None        Edits: None        Rung: Append 5/15 File DEMO

Branch            Append    Insert    Modify    Delete    Undel                  Accept
  *               Instr     Instr     Instr     Instr     Instr                  Rung
  F1              F3        F4        F5        F6        F7                      F10
```

• *SELECT BRANCH*   F1

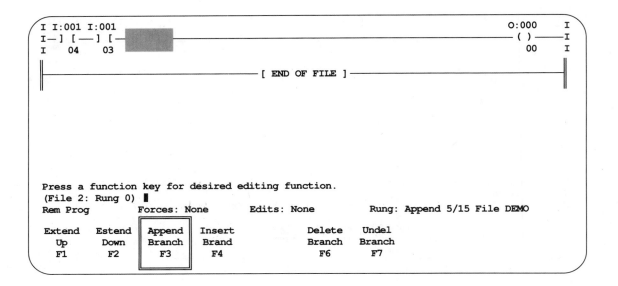

```
I I:001 I:001                                                      O:000    I
I—] [ —] [—░░░░░░░░░░░———————————————————————————————————( ) ———I
I   04    03                                                        00      I
 ||——————————————————————————[ END OF FILE ]——————————————————————||
 ||                                                                 ||

 Press a function key for desired editing function.
 (File 2: Rung 0) █
 Rem Prog          Forces: None        Edits: None        Rung: Append 5/15 File DEMO

 Extend    Estend   Append   Insert              Delete   Undel
  Up        Down    Branch   Brand               Branch   Branch
  F1        F2       F3       F4                   F6       F7
```

• *APPEND BRANCH*   F3

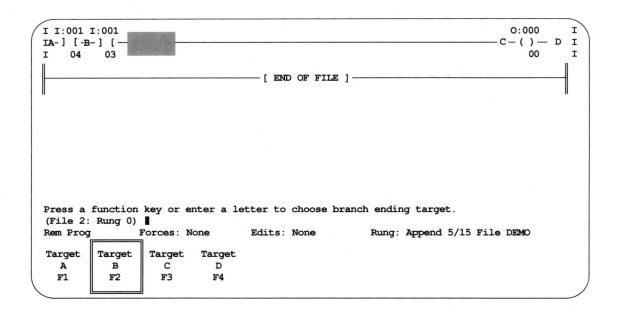

```
I I:001 I:001                                                      O:000    I
IA-] [ -B-] [—░░░░░░░░░░░—————————————————————————————C— ( ) — D  I
I   04    03                                                        00      I
 ||——————————————————————————[ END OF FILE ]——————————————————————||
 ||                                                                 ||

 Press a function key or enter a letter to choose branch ending target.
 (File 2: Rung 0) █
 Rem Prog          Forces: None        Edits: None        Rung: Append 5/15 File DEMO

 Target   Target   Target   Target
   A        B        C        D
  F1       F2       F3       F4
```

• Select target B   F2

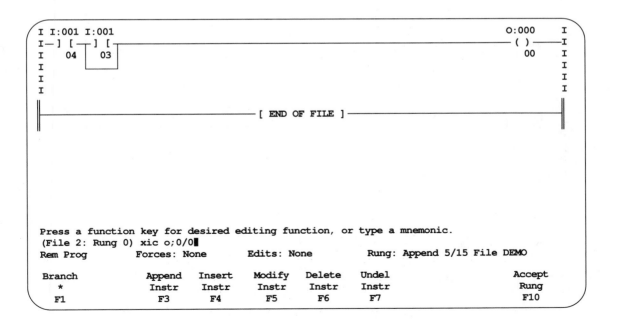

• Type instruction for holding contact   **xic o;0/0**

• Press *ENTER* key   *<ENTER>*

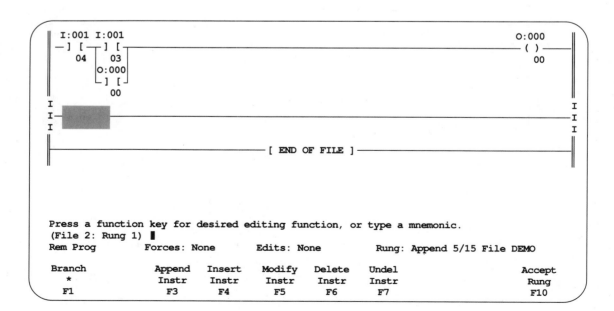

```
I I:001 I:001                                                    O:000    I
I—] [—┬—] [———————————————————————————————————————————————————( )———— I
I   04 │  03                                                      00     I
I      │O:000│                                                           I
I      └—] [—┘                                                           I
I         00                                                             I

‖——————————————————————[ END OF FILE ]————————————————————————————————‖

Press a function key for desired editing function, or type a mnemonic.
(File 2: Rung 0) ▌
Rem Prog          Forces: None        Edits: None          Rung: Append 5/15 File DEMO

Branch            Append   Insert    Modify   Delete   Undel              ┌─────────┐
   *              Instr    Instr     Instr    Instr    Instr              │ Accept  │
   F1             F3       F4        F5       F6       F7                 │ Rung    │
                                                                         │ F10     │
                                                                         └─────────┘
```

• Accept rung   F10

```
‖  I:001 I:001                                                   O:000   ‖
 —] [—┬—] [———————————————————————————————————————————————————( )———
    04 │  03                                                     00
       │O:000│
       └—] [—┘
          00
I                                                                       I
I—▓▓▓▓▓▓▓▓▓—————————————————————————————————————————————————————————————I
I                                                                       I
‖——————————————————————[ END OF FILE ]————————————————————————————————‖

Press a function key for desired editing function, or type a mnemonic.
(File 2: Rung 1) ▌
Rem Prog          Forces: None        Edits: None          Rung: Append 5/15 File DEMO

Branch            Append   Insert    Modify   Delete   Undel              Accept
   *              Instr    Instr     Instr    Instr    Instr              Rung
   F1             F3       F4        F5       F6       F7                  F10
```

• Complete program by hitting escape key twice   *ESC   ESC*

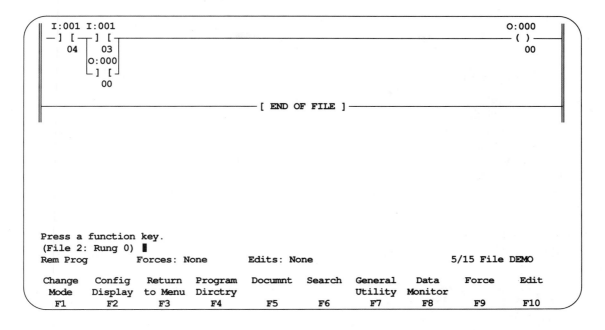

```
   I:001 I:001                                                    O:000
  —] [—┬—] [—┬                                                  —( )—
    04  │  03 │                                                    00
        │O:000│
        └—] [—┘
           00

  ├————————————————————————————[ END OF FILE ]————————————————————————————┤



  Press a function key.
  (File 2: Rung 0) ▮
  Rem Prog              Forces: None        Edits: None              5/15 File DEMO

  Change   Config   Return   Program   Documnt   Search   General   Data     Force    Edit
  Mode     Display  to Menu  Dirctry                       Utility  Monitor
  F1       F2       F3       F4        F5        F6       F7        F8       F9       F10
```

This method greatly reduces the number of keystrokes that are required and is much faster. It does, however, require that the programmer knows the mnemonic names of the various instructions, and be somewhat proficient at typing.

Peripherals

A communication port (or ports) is mounted on the back of the programmer, computer, or on the processor, for connection to a printer for hard copy printouts of the program, storage registers, or report generation. To use a printer, the printer must be compatible with the programmer or processor.

Some printers can be connected for **serial** and/or **parallel** communication. In serial communication, the bits are sent one after the other, or sequentially. In parallel communication, groups of bits (a byte) are sent simultaneously. When using a serial printer, the rate of communication (**baud rate**) between the printer and programmer varies with the printer and/or programmer capabilities. A baud rate of 110 indicates that 10-, 7- or 8-bit characters can be printed per second; 300 baud would be 30 characters per second; 1200 baud would be 120 characters per second, and so on.

Before purchasing a printer, be careful to check the communication baud rate. There is no sense in buying a printer capable of printing 120 characters per second (1200 baud rate) if the programmer can only communicate at 300 baud, or to purchase a printer that can only print 30 characters per second (300 baud rate) for a PLC or computer capable of 1200 baud. For printers that do not print as fast as the PLC or computer can communicate, an internal **buffer** is often used. A buffer is a temporary storage area where information can be held until the printer catches up.

Chapter Summary

The programming device (programmer) is used to enter, modify, and monitor the user program. The program (ladder diagram) is entered by pushing keys on the keyboard in a prescribed sequence so that the results can be displayed on either the VDT of a dedicated desktop programmer and the computer programmer, or on a LED or liquid crystal display on a hand-held programmer. The visual display can also be used as a troubleshooting aid to test the circuit prior to entry into user memory, or after the circuit is entered into memory and is operational. Contacts and coils are either intensified or displayed in reverse video to indicate true logic or power flow. From the programming device, contacts and coils can be forced *ON* or *OFF* while the circuit is operational. The FORCE ON, FORCE OFF capability should be *restricted* to personnel who have a *complete* understanding of the circuit and the driven equipment. Programming a PLC is not difficult, but time must be spent to become familiar with it and its programming techniques.

REVIEW QUESTIONS

1. What does the term *on line programming* mean?
2. What is the function of the cursor?
3. What is the FORCE feature used for?
4. T   F   Timer and counters use words of memory, but contacts, coils, and branch start instructions do not.
5. When is *off line programming* normally used?

Chapter 10 Programming Considerations

Objectives

After completing this chapter, you should have the knowledge to
- Define a *network*.
- Describe the term *dummy relay*.
- Understand the horizontal and vertical contact limits.
- Define the term *nesting*.
- Correctly wire and program *STOP* buttons.
- Describe the difference between logical and discrete holding contacts.

Network Limitations

A **network** is defined as a group of connected logic elements used to perform a specific function. Figure 10–1 shows a typical network consisting of seven series contacts and three parallel branches. A network also constitutes one rung of a ladder diagram that starts at the left rail and ends at the right rail.

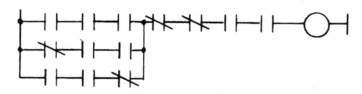

Figure 10–1 Network (Rung)

Some PLC manufacturers have virtually no network limitations, whereas other PLC systems are limited by the number of contacts or other logic symbols that can be included on the horizontal line of a network and the parallel branches (lines) that make up one network. A typical PLC network limitation of 10 series contacts per line and seven parallel lines, or branches, is shown in Figure 10–2. Additionally, some PLCs are further limited because they only allot one output per rung or network, and the output must be on the first line.

NOTE: Special functions and other logic symbols alter the network limitations and requirements; check the operation or program manual of the specific PLC for additional information.

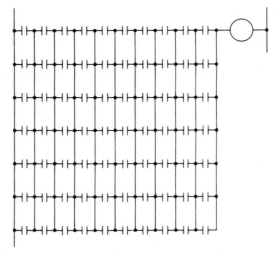

Figure 10–2 Network Limits

While the number of elements and lines within a network is limited, only the size of user memory limits the number of networks or rungs.

When a circuit requires more series contacts than the network allows (Figure 10–3a), the contacts are split into two rungs (Figure 10–3b). The first rung contains part of the required contacts and is programmed to an internal, or "dummy," relay. Internal relays are actually a bit and word location in storage memory or an unused bit in the I/O table.

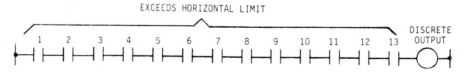

Figure 10–3a Contacts Exceed Horizontal Limit

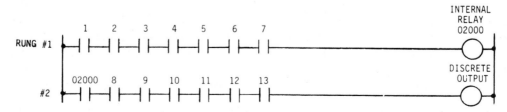

Figure 10–3b Contacts Split into Two Rungs

The address of the internal relay, 02000 in Figure 10–3b, is also the address of the first N.O. contact on the second rung. The remaining contacts (8–13) are programmed, followed by the address of the real-world output device. When the first seven contacts close, the internal output, bit 00 of word 020, is set to 1. This makes the N.O. contacts 02000 of Rung 2 true. If the other six contacts (8–13) are closed, the rung is true, and the discrete output is turned *ON*.

NOTE: It is not necessary to split the contacts in any ratio. If the network allows 10 horizontal contacts, 10 could be placed on the top rung, and three could follow the N.O. contacts of the internal relay (02000) on Rung 2. This technique applies not only to N.O. contacts, but also to N.C. (or combinations of N.O. and N.C.) contacts as well.

The internal or "dummy" relay just used does not exist as a real-world device that has to be hard-wired, but is merely a bit in the storage memory that performs the logic of a relay. In actual programming, internal control relays that do not actually exist, except in the storage memory as bits, are extensively used. The use of these internal relay equivalents is what makes the programmable controller unique, and eliminates hours of hard-wiring that shortens installation and maintenance time.

When a program requires more parallel branches than the network allows, the circuit can be split into two networks, or rungs. The first six parallel contacts are programmed to an internal or dummy relay as shown in Figure 10–4. A contact with the same address as the internal, or dummy, relay is then programmed in parallel with the remaining contacts to control the output.

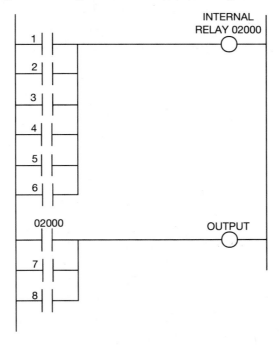

Figure 10–4 Parallel Contacts Split Into Two Rungs (Networks)

Sometimes it is not the PLC that limits the number of horizontal or parallel contacts, but the display capability of the VTD. The screen may only be able to display 10 horizontal contacts, even though the PLC can be programmed with an unlimited number of horizontal contacts. While the processor may allow more contacts to be programmed than the VDT can display, it is not a good idea to do so. If the contacts cannot be seen on the screen, it is difficult to troubleshoot the circuit later. The ability to view each contact on the VDT and to know the status (*ON* or *OFF*) of each contact, as well as the status of the output devices, is what makes the PLC such a powerful tool.

NOTE: When using a computer as a programming device, most software packages allow the electrician or technician to "shift" the screen to monitor all the instructions, even when the normal screen is filled with instructions.

Many PLCs have networks that allow for more than one output (parallel outputs) (Figure 10–5). With this parallel output configuration, all of the outputs are *ON* or *OFF* at the same time, based on the network logic.

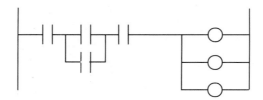

Figure 10–5 Parallel Outut Format

Other PLCs, like the Allen-Bradley PLC-5, allow multiple outputs that can be *ON* or *OFF* at different times, depending on the network logic (illustrated in Figure 10–6).

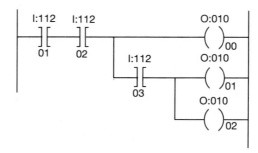

Figure 10–6 Multiple Outputs

Programming Restrictions

In addition to the number of horizontal contacts on one line, and the number of lines in a network or rung, the PLC does not allow for programming vertical contacts (Figure 10–7). In the real world, one could wire the circuit as shown in the figure, but programming restrictions would not allow the PLC to be programmed in this manner.

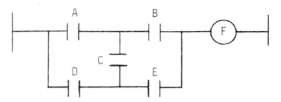

Figure 10–7 Vertical Contacts

If one analyzes the logic of the circuit in Figure 10–7, the circuit logic shows that output F can be energized by any of the following contact combinations: A, B (Figure 10–8a); A, C, E (Figure 10–8b); D, C, B (Figure 10–8c); and D, E (Figure 10-8d).

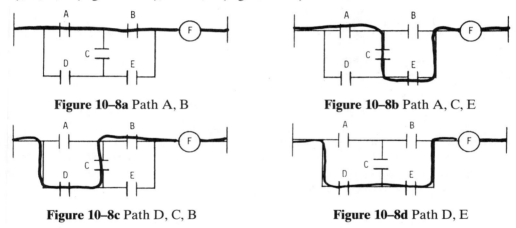

Figure 10–8a Path A, B   **Figure 10–8b** Path A, C, E

Figure 10–8c Path D, C, B   **Figure 10–8d** Path D, E

To duplicate the logic, the circuit could be programmed as shown in Figure 10–9.

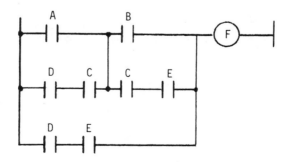

Figure 10–9 Equivalent Circuit without Vertical Contacts

This circuit maintains the circuit logic. Contact combinations A, B; A, C, E; D, C, B; and D, E all energize output F.

Another limitation to circuit programming is the way in which the processor considers power flow, or logic continuity, when it scans a rung of logic. Flow is from left to right *only*, and vertically *up* or *down*. The processor *never* allows logic continuity (power flow) from right to left.

Normally, relay logic for the circuit shown in Figure 10–10 would indicate the following possible contact combinations to energize output G: A, B, C; A, D, E; F, E; and F, D, B, C.

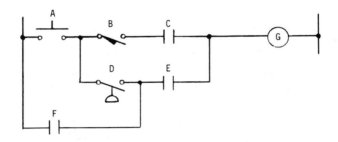

Figure 10–10 Hard-Wired Circuit

If the circuit shown in Figure 10–10 was programmed into user memory as shown in Figure 10–11a, the processor would ignore contact combination F, D, B, C because it would require power flow (logic continuity) from right to left. If combination F, D, B, C was required, the circuit would be reprogrammed as shown in Figure 10–11b.

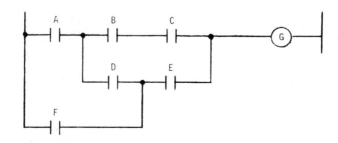

Figure 10–11a Circuit Improperly Programmed

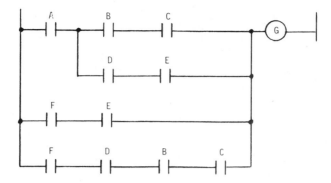

Figure 10–11b Circuit Properly Programmed

The last restriction placed on the programming of circuits into user memory by some—*but not all*—PLCs is the use of "a branch circuit within a branch circuit," or the **nesting** of contacts. Figure 10–12a is an example of a circuit that has nested contacts (L and G) or "a branch within a branch." To obtain the required logic, the circuit is programmed as shown in Figure 10–12b. The duplication of contacts J and K eliminates the nested contacts L and G.

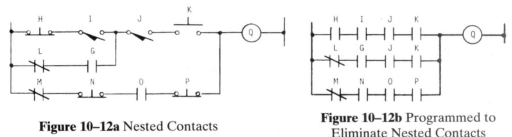

Figure 10–12a Nested Contacts

Figure 10–12b Programmed to Eliminate Nested Contacts

Figures 10–13a and 10–13b are other examples of circuits with "a branch within a branch" (in this case "branches within a branch"), and how the circuits are programmed to maintain circuit logic.

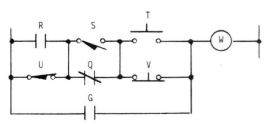

Figure 10–13a Branches Within a Branch

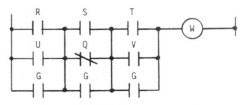

Figure 10–13b Programmed to Eliminate Branches Within a Branch

The easiest way to avoid nesting is to remember that all branches must start at a common point, and all of the branches must end at the same location. Figure 10–13b is actually three parallel branches in series as illustrated in Figure 10–14.

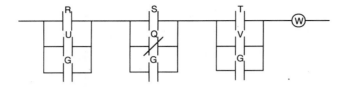

Figure 10–14 Parallel Series Combination

Program Scanning

As discussed in Chapter 3, the processor first determines the status of the input devices, then it scans the user program, and then updates (turns *ON* or *OFF*) the outputs. The way the processor scans the program varies from PLC to PLC. One common method is to scan the program from left to right and top to bottom, similar to the way in which a book is read. In this method, the processor scans the first rung of the program from left to right, then the second rung from left to right, and continues in this fashion until all the rungs have been scanned. In the next scan, the processor returns to the first rung and starts all over again, scanning each rung in order from top to bottom. Figure 10–15 illustrates the order of scanning for a processor that scans from left to right and top to bottom.

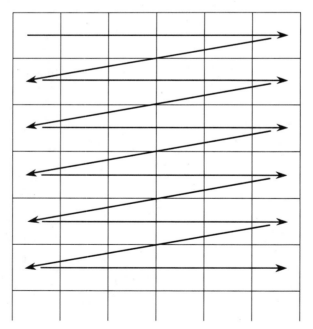

Figure 10–15 Processor Scan Left to Right, Top to Bottom

Look at the circuit shown in Figure 10–16. If S1 is closed, or true, the logic of Rung 1 is true, making the logic of Rung 2 true, which in turn makes the logic of Rung 3 true. Lamps 1, 2, and 3 are turned *ON* at the end of the first scan.

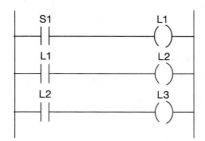

Figure 10–16 One Scan Turns On L3

If, however, the circuit was programmed as shown in Figure 10–17, L3 would *not* turn *ON* until the third scan was completed.

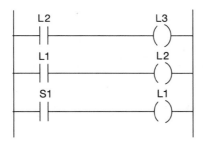

Figure 10–17 Three Scans Required to Turn On L3

In the first scan, Rung 1 is not yet logically true because the processor does not know the status of L2. Rung 2 would also not have logic continuity because the processor does not yet know the status of L1. When the processor scans Rung 3, however, it now is logically true because S1 is closed. The processor, therefore, turns *ON* L1 at the end of the first scan. On the second scan, L2 is still false, so Rung 1 has no logic continuity. Rung 2, however, is now logically true because L1 was turned *ON* after the first scan, and L1 contacts are closed, or true. S1 is still closed which keeps the third rung true, so at the end of the second scan, L1 and L2 are *ON*. It is only during the third scan that Rung 1 becomes true. With L2 now *ON*, the logic of Rung 1 is complete and L3 will be turned *ON* at the end of the third scan.

Each scan took only a matter of milliseconds (msecs) to complete, so the delay in turning on L3 is not perceptible to the naked eye. However, in certain high-speed processes, the time lost due to poor programming may be significant and must be considered.

Another example of how the processor scans is illustrated by duplicating output addresses. If the same output address is inadvertently used twice in one program, the last rung in which the address is used will indicate the status of the output. For example, if the address is used in Rung 5 and the logic of Rung 5 is true (which would tell the processor to turn *ON* the output), and the address is used again in Rung 13 and the logic of Rung 13 is false, the output will not be turned *ON* because the processor scanned Rung 13 last, and the last state of the output (true or false) will be based on the last rung scanned.

Programming STOP Buttons

During the early days of programmable controllers, it was common for salespeople to use a demonstrator model that had the *STOP* buttons wired normally open (N.O.). This technique was used so the salespeople did not have to explain why a *STOP* button was shown as normally open in a PLC program.

Figure 10–18a shows a standard *STOP/START* station ladder diagram, and Figure 10–18b shows the equivalent diagram used by some PLC salespeople.

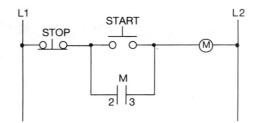

Figure 10–18a Standard *STOP/START* Ladder Diagram

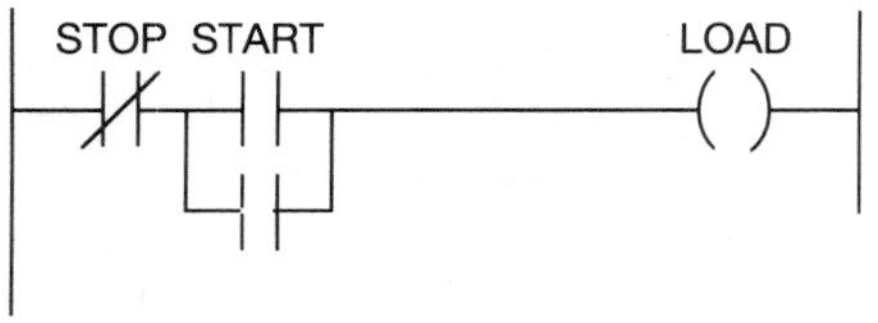

Figure 10–18B PLC Programmed *STOP/START* Circuit

From an understanding of how a *STOP/START* circuit works, and an understanding of the EXAMINE ON and EXAMINE OFF instructions, it is easy to see that the only way the circuit could be logically true would be for the *STOP* button to be wired open. By using a N.O. *STOP* button, the EXAMINE OFF instruction is true and the circuit energizes when the *START* button is pressed. The problem is that once the circuit is energized, the only way it can be stopped, or turned OFF, is if the *STOP* button is pushed and the contacts close. While this circuit will work, there is a built in *danger* that must be considered. If a *STOP* button is wired in a N.O. position, the switch is impossible to close if it becomes jammed, and it would be impossible to deenergize the circuit. Similarly, if a wire breaks on the *STOP* button, it is possible to complete the logic of the circuit and energize the equipment, but *impossible* to deenergize the equipment.

With the wire broken, changes in the status of the *STOP* button cannot be conveyed to the processor, and the circuit and/or equipment cannot be deenergized.

SAFETY NOTE: All *STOP* buttons must be wired so that a failure of the switch or a broken wire will automatically break logic continuity and turn the circuit *OFF*. A good programmer will always wire the devices and program the circuit so that if the real-world device fails, it creates a safe condition, not a safety hazard.

This practice of wiring *STOP* buttons in the N.O. position was common during the 1980s. If an electrician or technician finds equipment wired in this fashion, he or she should change the *STOP* buttons to N.C. and the PLC program from EXAMINE OFF to EXAMINE ON.

The *START* button should be wired normally open (N.O.) and programmed with an EXAMINE ON instruction. As a general rule, all input devices, except *STOP* buttons, are wired normally open and given EXAMINE ON instructions in the program.

Logical Holding Instructions

In previous programming examples we have used the output address to address the holding contacts. This method of providing holding logic works well in many applications and eliminates the need to actually wire the holding contacts on the motor starter. When the output address is also used for the holding contacts (logic), the circuit is maintained logically because the output point has been turned *ON*. This is no guarantee, however, that the actual motor starter connected to the output module has been energized.

Discrete Holding Contacts

The only real way to know that the starter has energized is to actually wire the holding contacts of the motor starter to an input module point, and use that address when programming the holding contacts. This method has many advantages, and short of installing a motor sensor, is the best way to verify that the motor starter has been energized.

Overload Contacts

It is common practice *not* to wire the overload contacts to an input module, but instead to wire the overload contacts in series with the starter coil (shown in Figure 10–19). When wired in this manner, a motor overload that opens the overload contacts also opens the circuit to the starter coil, and the starter will drop-out, or deenergize. When the starter deenergizes, the holding contacts (which must be wired to an input module and programmed in the PLC circuit) also open, and the PLC circuit loses logic, which in turn, turns *OFF* the point on the output module that is connected to the starter coil. This arrangement only works if the holding contacts are wired to an input module and programmed into the PLC program.

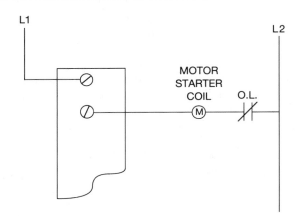

Figure 10–19 Overload Contacts Wired in Series
with the Starter Coil

If the holding contacts are not wired to an input module, but instead the output address is used for holding logic, the overloads trip and interrupt the circuit to the starter coil and the starter deenergizes, but the PLC logic is *not* broken. Without the holding contact to open the logic in the PLC program, the output module point would remain *ON*, even though the starter coil circuit is open. This wiring scheme can cause a safety hazard. With the logic remaining true, the motor restarts automatically when the overload is reset.

Not wiring the overloads to an input module and programming them in the circuit can also cause problems in sequential motor circuits, or other automated circuits that require one device to be *ON* before other devices are turned *ON*.

EXAMPLE: On a milling machine it is necessary for the coolant pump to always be running before the cutting wheel can operate. If the overload for the coolant pump is not wired to an input module, but wired in series with the motor starter coil instead, an overload opens the circuit to the coolant pump motor starter and turns *OFF* the coolant pump. The programmed circuit is not aware that the coolant pump has shut down, and allows the cutting wheel to continue to run. If the overload had been wired to an input module, the continuity of the circuit would have been broken when the overload opened, and both the coolant pump and cutting wheel would have been turned *OFF*.

When the decision is made not to wire the holding contacts on the starter, but instead to use the address of the output for logic continuity, the actual overload contacts *must* be wired to an input module and referenced in the PLC program.

Hard-wiring the overload contact(s) to an input module and addressing the contact(s) in the PLC program has the added benefit of being able to look at the status of the overload contacts to determine if they have tripped. Any address used in the PLC program can be viewed on the programmer screen to determine if the device is *ON* or *OFF*. By viewing the screen, it can be determined that the overload has tripped. Knowing why the motor starter deenergized is helpful when troubleshooting.

Chapter Summary

Each PLC has a maximum network size, or matrix, that limits the number of horizontal and vertical contacts for any one network, or rung. The only limitation to the number of rungs (networks) is memory size. Since the processor reads power flow (logic) from left to right *only*, and vertically either *up* or *down*, the logic of a relay circuit must be examined carefully to ensure that the logic is maintained when the circuit is programmed into the user memory. The program device does not allow contact to be programmed vertically, but the logic of a ladder diagram with vertical contacts may be duplicated by adding additional contacts. Depending on the PLC, contacts may or may not be programmed as a "branch-within-a-branch," or nested.

Consideration must be given to the way that the processor scans the program to eliminate unnecessary scans before a line or rung of logic goes true. *STOP* buttons must always be wired as normally closed and use an EXAMINE ON instruction to work correctly and safely. Holding contacts can be either logical or discrete (real world). The discrete method has the added advantage of verifying that the motor starter has indeed energized. Hard-wiring the overload contacts to an input module has the added benefit of being able to look at the status of the overload by means of a programming device to determine if they have tripped.

REVIEW QUESTIONS

1. Define the term *network*.
2. Draw and label a diagram to show how a network that requires 14 series contacts to control a discrete output can be programmed on a PLC that limits series logic elements to 10 on one line.
3. Draw a circuit with nested contacts.
4. Draw a circuit that retains the logic of Figure 10–A, which has a vertical contact (D), so the circuit could be programmed into a PLC.

5. Power flow, or logic, in a PLC is considered to be (check all correct answers):
   a. up to down only
   b. up or down only
   c. left to right
   d. right to left
   e. up or down and from left to right
   f. up, down, and from right to left
   g. up to down only and left to right
   h. up to down only and right to left
   i. up or down and left to right or right to left
6. Write a program for a PLC that does *not* allow nested contacts, for the Hand-Off-Auto circuit shown in Figure 10–B.

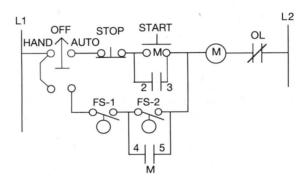

7. Explain how a *STOP* button must always be wired, and why.
8. List two ways that holding contacts can be programmed.
   a. _____
   b. _____
9. Explain one advantage of wiring the overload contacts to an input module and then programming the overload address into the PLC program.

Chapter 11

Program Control Instructions

Objectives

After completing this chapter, you should have the knowledge to
- Write a program using a *latching relay*.
- Understand the term *retentive*.
- Write a program using a *master control relay*.
- Understand the importance of a *safety circuit*.
- Describe how an *immediate input instruction* could be used.
- Describe how an *immediate output instruction* could be used.
- Write a program using the *jump and label instructions*.
- Give a reason for using the *temporary end instruction*.

Master Control Relay Instructions

In standard relay control systems, a master control is often used to control power to the entire circuit or just to selected rungs. This allows selected rungs, or the whole circuit, to be deenergized by turning off the **master control relay** (**MCR**). Figure 11–1 shows a typical hard-wired master control relay that controls power for the whole circuit.

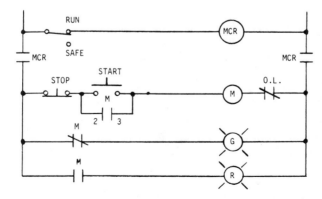

Figure 11–1 Hard-Wired Master Control Relay

Master control relays are often used with circuits that have off delay timers so the circuit can be shut down completely *without* waiting for the timers to time out.

With PLCs, an MCR function can be programmed to control an entire circuit or just selected rungs of a circuit. When the MCR instruction is programmed (as shown in Figure 11–2), any rungs that follow can only energize if the MCR is energized, or set to a true condition. When an MCR instruction is true, the outputs in the rungs that follow are controlled in a normal fashion by the logic programmed for each rung. If the MCR is deenergized, or reset to false, the rungs below the MCR are also deenergized, and *cannot* energize even if the programmed logic for each rung is true. The exception is **latching relay instructions**. Latching relay instructions remain *ON*, or true, even when the MCR instruction is false. Latching relay instructions are covered later in this chapter.

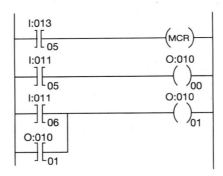

Figure 11–2 Allen-Bradley PLC-5
Programmed MCR

Figure 11–3 shows a programmed circuit with two MCRs (Square D format). By adding the second MCR, the rungs between the two MCRs—Rungs 3 and 4—are controlled by the first MCR and contact 03-01. The second MCR—Rung 5—is independent of the first and controls Rung 6 below it.

NOTE: MCRs are sometimes called **master control resets** because they reset the output to zero, or *OFF*.

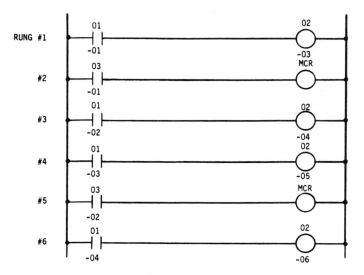

Figure 11–3 Using Multiple MCRs

To end the control of one MCR, a second MCR can be programmed *unconditionally*, with no contacts or logic preceding the MCR, as shown in Figure 11–4 (Allen-Bradley format). Any additional rungs below the second MCR will function normally with the outputs being energized when the programmed logic is true, regardless of whether the first MCR is energized or not.

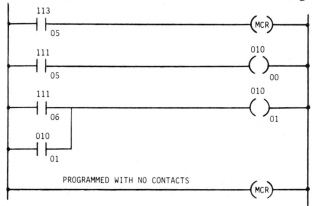

Figure 11–4 Unconditionally Programmed MCR

Latching Relay Instructions

Before discussing how a latching relay function is programmed with a PLC, a review of the traditional hard-wired latching relays may be helpful.

Latching relays are used when it is necessary for contacts to stay open and/or closed even though the coil is only energized for a short time (40 milliseconds) by a momentary signal. Latching relays are often used for lighting applications where multiple circuits are needed for the lights in a large room or auditorium. Instead of having a switch for each circuit, a multiple contact latching relay is used. Each lighting circuit is wired to a set of contacts on the latching relay. One switch now controls the latching relay, which in turn controls several circuits of the lighting load. The latch and unlatch feature of the relay only requires three wires, so wiring is greatly reduced, and contolling the latching relay from multiple locations is quite simple. Another advantage of using the latching relay occurs when lighting is normally turned *ON* at the start of the work day, and left *ON* until quitting time. Under these circumstances, if a normal relay is used, the coil needs to be energized all day long to keep the lights *ON*. The latching relay, however, needs only to be momentarily energized to latch, or turn the connected load *ON*, and the relay remains latched even though the coil is no longer energized. By *not* having the coil energized all day, there is an energy saving.

Latching relays normally use two coils: one to *latch* and one to *unlatch*. There is a mechanical linkage that holds the relay in the latched, or closed, position. When the unlatch coil is energized, the coil action disengages the mechanical latch and allows the relay to open.

Figure 11–5 shows the wiring diagram for a mechanical latching relay.

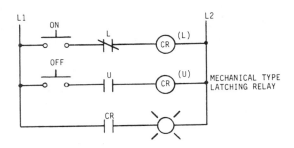

Figure 11–5 Mechanical Latching Relay

When the *ON* button is pushed, the latch coil energizes and opens the N.C. latch (L) contacts and closes the unlatched (U) contacts. Opening the N.C. L contacts deenergizes the L coil. The length of time it took to push the *ON* button and to energize the latch coil (which opened the N.C. L contacts and deenergizes the L coil) required only a fraction of a second. During the short time the latch coil energized, it closed the N.O. CR contacts, completing the circuit to the lamp. CR contacts remain closed even though the latch coil deenergizes because of the mechanical latch mechanism. To open the mechanically latched contacts to turn the light *OFF* requires the *OFF* push button be pushed. The U contacts in the unlatch coil circuit are now closed. Pushing the *OFF* button energizes the unlatch coil, which in turn closes the N.C. L contacts, and opens the U contacts, which deenergizes the unlatch coil. For the brief instant that the unlatch coil is energized, it releases the mechanically latched CR contacts so they can open and turn the light *OFF*.

Mechanical latching relays can be replaced by programming bits in the storage memory of the PLC as internal latching instructions, or by programming real-world outputs using the latch and unlatch instructions. Like the dummy relays discussed earlier, the programmed internal latching relays do not exist as real-world devices but can perform all the logic of an actual latching relay.

Programmed latch and unlatch instructions, like their physical real-world counterparts, are **retentive** during a power failure. For example, if the relay is latched, it remains latched if power is lost and then restored; or if modes are changed from *run* to *program*, and then back to *run*, the output remains in its last state (latched).

Figure 11–6 shows latch and unlatch Rungs as they are programmed on an Allen-Bradley PLC-2/30.

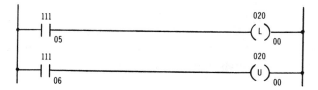

Figure 11–6 Programmed Latch and Unlatch Rungs

Both the latch L and unlatch U coils have the same address (02000). The address (bit 00 of word 020) is set to 1, or *ON*, when the latch Rung is true (input 11105 closed), and will be cleared to 0, or *OFF*, when the unlatch Rung is true, (input 11106 closed). Similar to normal latching relays, only a momentary closure of input device 11105 latches the output coil (L) 02000, and the output will remain latched or *ON* until the unlatch coil (U) Rung is true by closing input 11106.

N.O. and N.C. contacts programmed after the latch and unlatch Rungs, and given the same address as the latch and unlatch coils, perform the same functions as N.O. and N.C. contacts of a standard real-world latching relay (Figure 11–7).

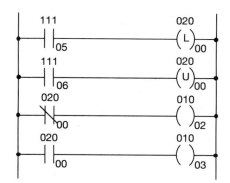

Figure 11–7 Programmed
Latching Circuit

Output 01002 is *ON* and output 01003 is *OFF* when 02000 is *not* latched.

Output 01002 goes *OFF* and 01003 comes *ON* when 02000 *is* latched.

Normally, an internal storage bit (dummy relay) is used for the latch and unlatch address, rather than an actual discrete output address. If a discrete output address is used, the output, once latched, remains *ON*, even if programmed after an open MCR Rung. When an internal storage bit is used for the latch and unlatch address, the bit is still retentive, but turns *OFF* if programmed *after* an MCR Rung that is open.

The Allen-Bradley PLC-5s use an output latch (OTL) instruction and output unlatch (OTU) instruction for latching relay functions. The OTL instruction is retentive. When the processor loses power, is switched to either the *test* or *program* modes, or detects a major fault, outputs are turned *OFF*; the state of the OTL is retained in memory, however, and when power is restored or the processor is switched back to the *run* mode, the outputs that were *ON* previously return to their *ON* state. Figure 11–8 shows how the PLC-5 OTL and OTU appear after they have been programmed. Notice that the format is the same as it is for the PLC-2 family, except for the address configuration.

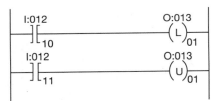

Figure 11–8 PLC-5 Latch and Unlatch Format

Although all PLCs are designed and manufactured to the highest standards and quality, a latching or MCR instruction should not be depended on for machine safety. A hard-wired **safety circuit** should always be added. A safety circuit is recommended by most PLC manufacturers to ensure maximum safety rather than depending on a programmed MCR or latching relay alone.

Safety Circuit

The concept of a safety circuit has been discussed earlier in the text, and is an important enough subject to be covered again. The National Electrical Manufacturing Association (NEMA) standards for programmable controllers recommends that consideration be given to the use of **emergency stop** functions which are independent of the programmable controller. The standard reads in part:

When the operator is exposed to the machinery, such as loading or unloading a machine tool, or where the machine cycles automatically, consideration should be given to the use of an electromechanical override or other redundant means, independent of the controller, for starting or interrupting the cycle.

Figure 11–9 shows how a control relay (CR) and *safe-run switch* is added to interrupt line 2 to the discrete output devices of an automatic machine or process.

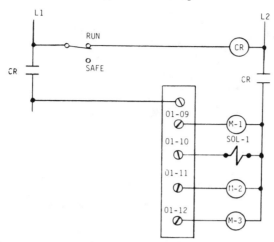

Figure 11–9 Safety Circuit

Immediate Input Instruction

The **immediate input instruction** is used when it becomes necessary to update the status of an input word in the input image table prior to the normal scan sequence. For some high-speed processes it may be necessary to know the status of a certain bit (or bits) in a particular input word prior to the next normal input image table update. Allen-Bradley refers to their immediate input instruction with the mnemonic IIN. To use the instruction to interrupt the normal scan cycle and update an input word, the IIN instruction is programmed prior to the rung(s) that include critical input addresses. The address for the IIN instruction is the rack number and module group for the input devices that need to be updated prior to the normal update. The rack number and the module group correspond to a word in the input image table. Figure 11–10 illustrates how the IIN instruction is programmed.

Figure 11–10 Programming an Allen-Bradley
PLC-5 Immediate Input Instruction

If input I:010/00 is true, the program scan is interrupted while the input status for input image word 12 is updated (rack 1, module group 2). As soon as the input word has been updated, the scan continues. Figure 11–11 shows the typical scan cycle and how it is interrupted for the immediate update of the status of the input devices in rack 1, module group 2.

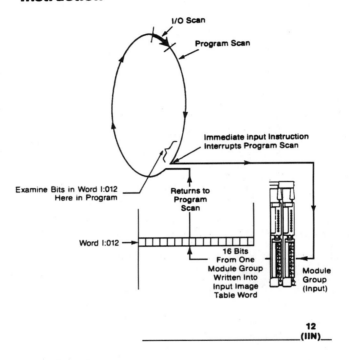

Immediate Input Instruction

1. Interrupt program scan
2. Update input image table word with current status of associated I/O group
3. Program scan continues
4. Program uses new input status

Figure 11–11 Scan Interrupted for Immediate Input Update
(Courtesy of Allen-Bradley)

Immediate Output Instruction

The **immediate output instruction** is used when it is necessary to update the output image table prior to the completion of a normal program scan. Allen-Bradley PLC-5 uses the mnemonic IOT. When the immediate output instruction (IOT) is enabled, or true, the instruction will interrupt the normal program scan and update a word in the output image table. Similar to the IIN instruction, the rack number and module group number is used to identify the location of the output devices that are to be immediately updated. Figure 11–12 shows how the instruction is programmed for outputs in rack 1, module group 3 (output word 013).

```
    I:010                              13
——] [——————————————————————————————(IOT)——
    00
```

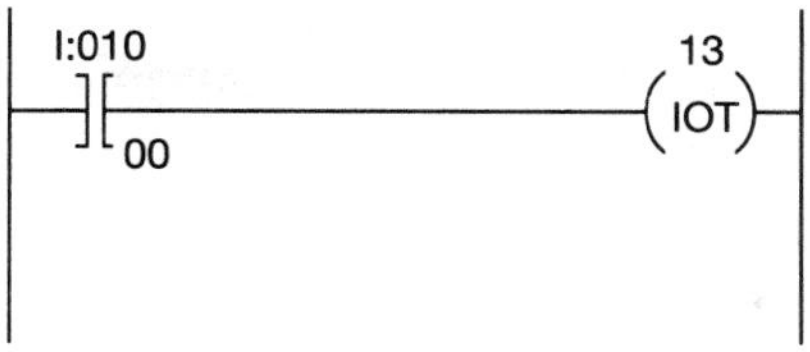

Figure 11–12 Programming the Immediate Output Instruction

When input I:010/00 is true, the IOT instruction is enabled and the scan is interrupted so that the output devices in rack 1, module group 3 (output word 013) are immediately updated. Figure 11–13 shows the interrupted program scan for the IOT instruction.

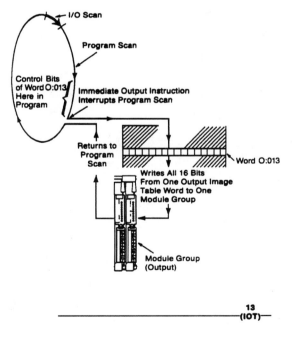

Immediate Output Instruction

1. Control output bits for outputs of interest
2. Interrupt program scan
3. Send status of output image table word to associated output module(s)
4. Continue program scan

Figure 11–13 Scan Interrupted for Immediate Output Update *(Courtesy of Allen-Bradley)*

Jump and Label Instructions

Used in combination, these two instructions allow for skipping over portions of the program to save program scan time. If there is a portion of the program that is not operational during certain portions of the process, the portion that is not used and/or needed can be jumped over or by-passed until it is needed again. By jumping over parts of the program, the scan time is decreased and more scans can be completed in a given period of time, which, in turn, means more frequent updating of information in the program. The **jump instruction (JMP)** tells the processor to jump over a portion of the program. Where to jump to is controlled by the **label instruction (LBL)**. Figure 11–14 shows a jump and label instruction in Allen-Bradley PLC-5 format. When the JMP instruction is true, the processor will jump over Rungs 3 and 4 and go directly to Rung 5, as shown.

The JMP instruction is assigned a three-digit number from 000–255, and the rung that the processor is to jump to is given an LBL of the same number. In Figure 11–14, the JMP and LBL instructions are given the number 20. When input address I:012/13 is enabled, or true, the JMP instruction is also enabled, and the processor is instructed to jump all successive rungs until it reaches the rung that contains the label instruction with the number 20. In this illustration, only two rungs are jumped. In actual practice, any number of rungs can be jumped.

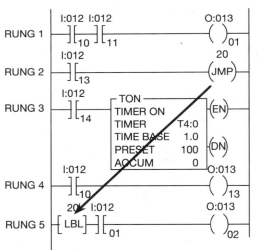

Figure 11–14 Programming the Jump (JMP) and Label (LBL) Instructions

The jump and label instruction can be used to jump forward or backwards in the program, depending on need. Jumping backward adds to the total scan time.

NOTE: Jumping backward an excessive number of times could increase the scan time to a point where the watchdog timer will time out (the processor has a watchdog timer that is reset on each scan). If the scan time exceeds the watchdog timer's preset time, the processor goes into a *fault* condition.

Jump to Subroutine, Subroutine, and Return Instructions

The **jump to subroutine**, **subroutine**, and **return instructions** are used to direct the processor to a subroutine file within the program, scan it, return to the program, and continue to scan. The formats used by the various PLC manufacturers to program these instructions varies widely, and for this reason, only instruction blocks are shown in Figure 11–15. The blocks are identified using the Allen-Bradley mnemonics, or labels: jump to subroutine (JSR); subroutine (SBR); and return (RET). Subroutines are very valuable for program organization, and for using blocks of programming logic over and over by simply changing the variables used in the subroutine. To use this group of instructions, consult the programming guide for the PLC system that is being used.

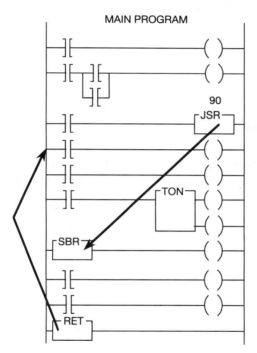

Figure 11–15 Jump to Subroutine, Subroutine, and Return Instructions

Temporary End Instruction

The **temporary end instruction (TND)** is used to place a temporary end to the program scan. When the instruction is inserted into the program, the processor stops scanning the program at this point and returns to the start of the program. When the processor returns to the start of the program, the watchdog timer is reset to zero. This instruction is often used when a new program is being debugged for the first time, because it allows for portions of the program to be checked out without running the entire program. Figure 11–16 shows a TND using the Allen-Bradley format. When the TND instruction is true, the processor stops scanning at Rung 3 and does not scan Rungs 4 and 5.

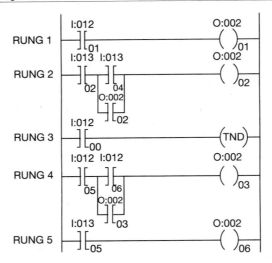

Figure 11–16 Allen-Bradley PLC—Temporary End Instruction (TND)

Always False Instruction

The **always false instruction (AFI)** is also used when debugging a new or modified program. By inserting the always false instruction in a rung, the rung will always be false, regardless of the status of other instructions in the rung. Figure 11–17 shows a rung of logic with the AFI instruction programmed at the start of the rung.

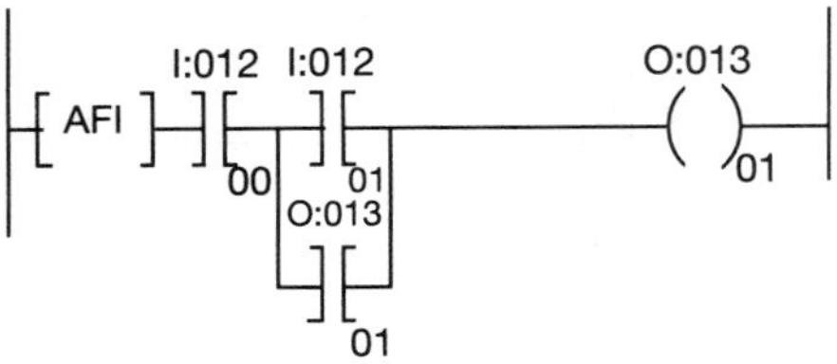

Figure 11–17 Always False Instruction

One Shot Instruction

The **one shot instruction (ONS)** is an input instruction that makes the rung true for just one program scan, based on a false-to-true transition of the instruction that precedes the one shot instruction. Figure 11–18 shows a rung of logic with the one shot instruction programmed after input device I:011/04, which is controlling output O:010/12. Again, the Allen-Bradley PLC-5 format is used.

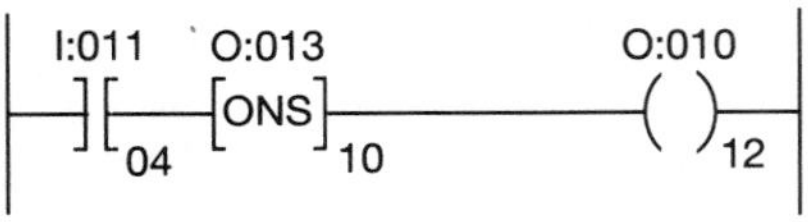

Figure 11–18 One Shot Instruction

When the rung is programmed as shown, the output is turned *ON* for one scan, and one scan *only* when input I:011/04 closes (makes a false-to-true transition). The output cannot be turned *ON* again until the input device is first opened, then closed again, making a false-to-true transition. With the next false-to-true transition, the output device is again only turned on for one scan. This is a beneficial instruction when an output signal or operation is wanted for only one scan.

Math operations, data or word moves, and the like, are completed only once if a one shot instruction is put in series or "anded" with the instruction. The one shot instruction is often used with timers and counters for changing preset and accumulated values. This technique is discussed further in Chapter 14.

NOTE: This chapter has covered some of the basic instructions that are available for programming with a PLC. As the instruction sets vary with each manufacturer, it is necessary that the programming manuals be consulted to determine what instructions are available, what their mnemonics, or designations are, and how to properly use them in a program.

Chapter Summary

Latching and master control instructions can be programmed to serve the same control functions as their real-world counterparts. Where personnel safety is a factor, a safety circuit should be added, instead of depending on latching or master control relay instructions alone. Through the use of various PLC instructions, the programmer can cause the processor to immediately update specific input and output devices, label and jump between specific rungs of logic, jump to subroutines then back to the original rung by using special instruction blocks, place a temporary end statement in the program to limit the amount of program that the processor will scan, use an always false instruction to keep a rung of logic from going true, and program one shot instructions that limit activity to only that one program scan. Although the mnemonics used by the various manufacturers will differ for each of their instruction sets or blocks, the main purpose of each instruction is the same. Once the electrician or technician has mastered programming one type of PLC, the transition to other types becomes easier.

REVIEW QUESTIONS

1. Will both programmed and real-world latching relays, if latched, remain latched if power is lost and then restored?
2. Latching relays are normally used when it is necessary for:
   a. contacts to open and/or close only while the coil is energized.
   b. contacts to open and/or close every 30 seconds.
   c. contacts to stay open and/or closed even though the coil is only energized a short time.
   d. none of the above.
3. Explain why a *master control relay* is often used with off delay timers.
4. A master control relay can be used to control:
   a. selected circuit rungs (networks).
   b. entire circuits.
   c. individual contacts within a rung (network).
   d. all of the above.

5. Define the term *unconditional*.
6. When using PLCs, the National Electrical Manufacturing Association (NEMA) recommends that consideration be given to stop functions independent of the PLC. Explain briefly why this recommendation is made.
7. Define the term *retentive*.
8. Give an example of how an *immediate input instruction* is used.
9. What is the primary function of an *immediate output instruction*?
10. What does the *jump* and *label instruction* do?
11. Give one reason why you might use a *temporary end instruction*.
12. What is the function of the *always false instruction*?
13. What is the function of a *one shot instruction*?

Chapter 12 Programming Timers

Objectives

After completing this chapter, you should have the knowledge to

- Describe how *pneumatic time delay relays* work.
- Write a program using *ON delay* and *OFF delay* timers.
- Describe the difference between an *ON delay timer* and a *retentive timer*.
- Explain how to extend the time range of timers by *cascading*.

Pneumatic Timers (General)

To fully understand how a PLC can be programmed to replace pneumatic time delay relays, both the basic pneumatic time delay relay and the standard symbols used must be understood.

Figure 12–1 shows a complete Allen-Bradley pneumatic timing relay, and Figure 12–2 shows a cutaway view of the contact and timing mechanism.

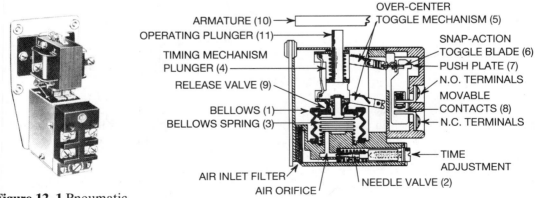

Figure 12–1 Pneumatic Timing Relay
(Courtesy of Allen-Bradley)

Figure 12–2 Cutaway View of Contact Unit and Timing Mechanism
(Courtesy of Allen-Bradley)

For the timer to time when power is applied (coil energized), the solenoid unit—coil, core piece, and armature—is mounted so that the natural weight of the armature (10) pushes down on the operating plunger (11). This causes the bellows (1) and bellows spring (3) to collapse the bellows, and dispel the air out through the release valve (9). When the coil is energized, the armature is attracted magnetically to the pole pieces, and lifts up and off the bellows assembly. Air now comes in through the air inlet filter, passed the needle valve (2), and fills the bellows with air. The incoming air expands the bellows upward, pushing on the timing mechanism plunger (4). As the plunger rises, it causes the over-center toggle mechanism (5) to move the snap-action toggle blade (6) upward. This picks up the push plate (7) that carries the movable contacts (8) to open the N.C. contact and close the N.O. contact. The time it takes for the bellows to fill with air and activate the contact mechanism is controlled by adjusting the needle valve in the air orifice. The valve is adjusted with a screwdriver as shown in Figure 12–3. A counterclockwise rotation moves the needle valve further into the air orifice, restricting airflow into the bellows, slowing the airflow, and increasing the time it takes for the bellows to expand and operate the contact mechanism. Conversely, clockwise adjustment of the needle valve decreases the time it takes the bellows to fill with air and activate the contacts after the armature has been lifted off the bellows mechanism.

Figure 12–3 Pneumatic Timer Adjustment
(Courtesy of Allen-Bradley)

When the contact action is delayed after the coil has been energized and the armature is lifted up and off the bellows mechanism, it is called **ON delay**.

When the coil of an ON delay timer is deenergized, the armature drops down, pushing on the operating plunger, which in turn pushes down on the bellows expelling air through the release valve. The downward motion of the bellows causes the snap-action toggle blade to instantaneously snap the N.C. contact closed and the N.O. contact open.

To summarize the ON delay timer, the delay in contact operation begins *after* the timer coil has been energized, or turned *ON*. When the timer coil is deenergized, or turned *OFF*, the contacts go back to their normal condition instantly. Figure 12–4 shows a pneumatic timer with the solenoid unit mounted for ON delay.

Figure 12–4 ON Delay Timer
(Courtesy of Allen-Bradley)

Figure 12–5 illustrates the electrical symbols used to indicate ON delay contacts.

The arrowhead indicates that movement is up. Since ON delay contacts can only time *after* the armature has lifted up off the bellows, this method of identifying timed contacts is easy to remember. Another common method of identifying timed contacts is shown in Figure 12–6.

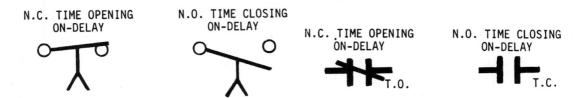

N.C. TIME OPENING ON-DELAY

N.O. TIME CLOSING ON-DELAY

N.C. TIME OPENING ON-DELAY
T.O.

N.O. TIME CLOSING ON-DELAY
T.C.

Figure 12–5 ON Delay Symbols   **Figure 12–6** ON Delay Symbols

NOTE: Remember that normal for contacts is how they are open or closed, with the coil of the relay deenergized and time expired.

For a pneumatic timer to time when power is removed from the relay coil (OFF delay), the solenoid unit is mounted as shown in Figure 12–7. With a spring holding the armature up, no weight is applied to the bellows assembly, and the bellows are filled with air in a fully extended position.

Figure 12–7 OFF Delay Timer
(Courtesy of Allen-Bradley)

NOTE: Compare Figure 12–4 (the ON delay) with Figure 12–7 (the OFF delay) to clearly see the difference in the mounting of the solenoid assemblies.

When the coil is energized and the armature moves down, the armature pushes on the operating plunger. The plunger pushes on the bellows assembly, and all air is immediately forced out of the bellows through the release valve. This causes the snap-action contact assembly to instantly open the N.C. contact and close the N.O. contact. The contacts stay in this configuration as long as the coil is energized and the armature is holding the bellows mechanism down (compressed).

When the relay coil is deenergized, or turned *OFF*, the spring on the armature lifts it up and off the operating plunger, which allows the bellows to fill with air. The N.C. contact remains open, and the N.O. contact remains closed until the bellows are filled with enough air to activate the snap-action contact mechanism. When the contact mechanism has been activated, the N.C. contacts go closed and the N.O. contacts go open.

Figure 12–8 shows the electrical symbols for OFF delay contacts.

To avoid confusion when reading electrical drawings with OFF delay contacts, it must be remembered that normal refers to the coil *after* it has been deenergized (turned *OFF*), and the time set for the timer has elapsed. The other symbols used for OFF delay contacts are shown in Figure 12–9.

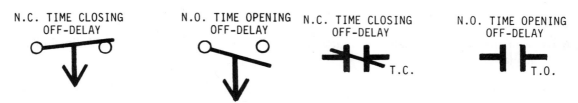

Figure 12–8 OFF Delay Symbols   **Figure 12–9** OFF Delay Symbols

Figure 12–10 compares both types of symbols used for ON delay and OFF delay timer relays.

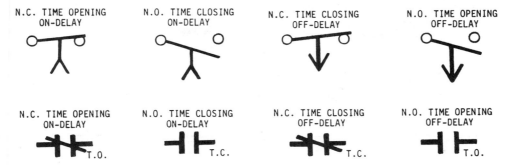

Figure 12–10 ON and OFF Delay Symbols

Reviewing the two types of symbols commonly used in motor control diagrams, an electrician or technician should have no trouble determining the type of timing relay (ON delay or OFF delay) used, or what is normal (open or closed) for the timed contacts.

The basic pneumatic timing relay is designed so that additional instantaneous contacts may be added, as shown in Figure 12–11. The instantaneous contacts operate when the coil is energized or deenergized independent of the timing mechanism. Figure 12–12 shows the electrical symbol for contacts with an asterisk (*) which is sometimes used to indicate instantaneous contacts of a timing relay.

N.O. INSTANTANEOUS CONTACTS
TIME DELAY RELAY

N.C. INSTANTANEOUS CONTACTS
TIME DELAY RELAY

Figure 12–12 Instantaneous Contact Symbols

Figure 12–11 Adding Instantaneous Contacts
(Courtesy of Allen-Bradley)

Figure 12–13a shows a simple light circuit controlled by an ON delay timer set for five seconds. The amount of delay is written near the timer coil on the diagram for understanding and for troubleshooting. Figure 12–13b shows that when S^1 is closed, the coil of the pneumatic timer energizes, lifts the armature up and off the bellows, and the timing starts. Figure 12–13c shows the circuit after three seconds have elapsed (not enough time for the timer to time out) with the lamp circuit still open. After five seconds have elapsed (Figure 12–13d), the N.O. time closing contacts close, and the lamp lights. As long as S^1 remains closed, the timer coil is energized, and the timed contacts stay closed. When S[1] is opened (Figure 12–13e), the coil circuit is broken, and the coil deenergizes. This causes the timed contacts to open, thereby turning OFF the lamp. The timed contacts will open the instant the coil deenergizes because they are timed only when power is applied to the coil.

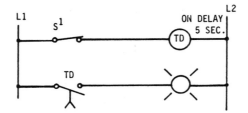

Figure 12–13a ON Delay Timer Circuit

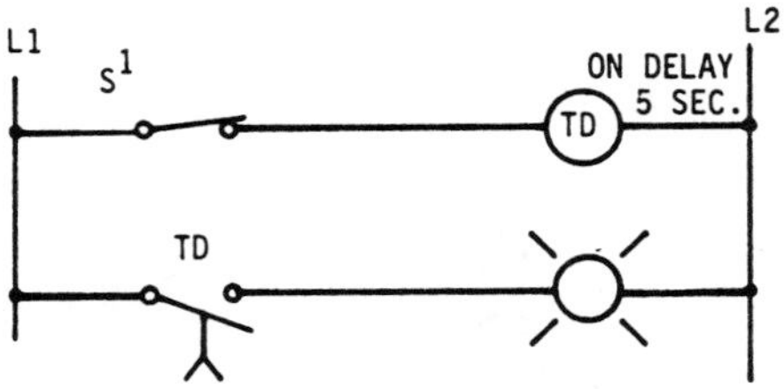

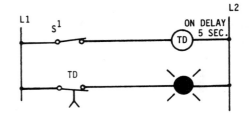

Figure 12–13b Instant S[1] is Closed

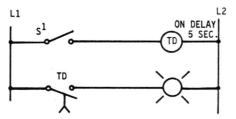

Figure 12–13c Three Seconds After S[1] is Closed

Figure 12–13d Five Seconds After S[1] is Closed

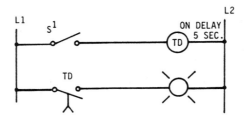

Figure 12–13e Instant S[1] is Opened

Figure 12–14a shows the same circuit but with an **OFF delay** timer. When S[1] is closed (Figure 12–14b), the TD coil energizes, drawing the armature down and compressing the bellows. This causes the N.O. OFF delay contacts to go closed instantly, and the lamp lights. When S[1] is opened (Figure 12–14c), the TD coil is deenergized, the spring-loaded armature is lifted up and off the bellows, and the five-second timing begins. Figure 12–14d shows the circuit after three seconds have elapsed. The lamp remains energized until the full five seconds have elapsed, and the N.O. contacts time out and open (Figure 12–14e).

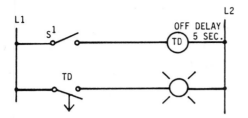

Figure 12–14a OFF Delay Timer Circuit

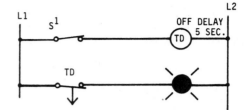

Figure 12–14b Instant S[1] is Closed

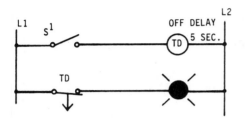

Figure 12–14c Instant S[1] is Opened

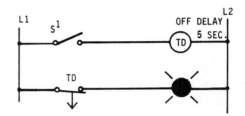

Figure 12–14d Three Seconds After S[1] is Opened

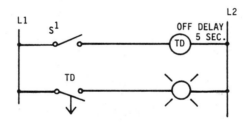

Figure 12–14e Five Seconds After S[1] is Opened

Instead of the bellows assembly, like the pneumatic time delay relay, PLC timers use internal solid state circuitry (clocks) for timing intervals or **time base**.

The various PLC manufacturers use varying approaches for the actual programming of timers. Several methods which are typical for most PLCs will be discussed.

Because it is an easy transition from pneumatic timer concepts to programming concepts, the Allen-Bradley PLC-2 approach to programming timers is discussed first.

Allen-Bradley PLC-2 Timers

The amount of time for which a timer is set is called the **preset time (PR)**. As the timer is activated and starts timing, time is **accumulated** until the preset time is reached. When the accumulated time *equals* the preset time, the contacts are activated.

Allen-Bradley PLC-2 timers use two words of memory: one word stores the preset time; the other word stores the accumulated time. When a timer is programmed, the operator must enter an address, a time base, and a preset time (PR) as indicated in Figure 12–15.

Figure 12–15 Programming an Allen-Bradley PLC-2 ON Delay (TON) Timer

The accumulated time (AC) displays the actual accumulating time as the timer is timing. The timer address is used to identify the first word that is used for the timer from the data table. On a standard 128-word data table, the first word set aside for timers and counters is word 030. If word 030 is used for the timer address, the first 12 bits, bits 00–07 and 10–13, of word 030 stores the accumulated time of timer 030 in a Binary Coded Decimal (BCD) format (Figure 12–16).

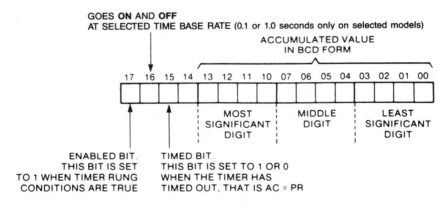

Figure 12–16 Time Counted Using BCD (12 bits)

NOTE: Allen-Bradley PLC-2 memory words and bit addresses use the octal numbering system.

Three of the last 4 bits are used as status bits. Bit 15 is set to 1, or 0, depending on whether the timer is **ON** or **OFF delay** (**TON** or **TOF**). Bit 16 on some systems goes *ON* and *OFF* (1 and 0) at the rate of the time base. Bit 17 is set to 1, or turned *ON*, whenever the timer is energized. Bit 15 acts like timed contacts, and bit 17 acts like instantaneous contacts.

As discussed in Chapter 5, the Allen-Bradley PLC-2 automatically sets aside the word numbered 100 higher than the timer for storing the preset value. If word 030 is addressed as a timer, word 130 is automatically used to store the preset value. Addressing word 031 as a timer would cause the preset value to be stored in word 131.

The time base can be seconds, tenths of a second, or hundredths of a second. Time bases are selected by entering numeric information into the processor by means of the programming device. For a time base of seconds, 10 is entered, and 1.0 will be displayed on the VDT. To set the time base for tenths of a second, 01 is entered, and the VDT displays 0.1. Entering 00 gives hundredths of a second, or 0.01 (Figure 12–17).

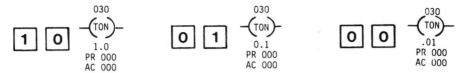

Figure 12–17 Time Bases

Next, the preset value is entered. The processor converts the decimal number(s) entered into BCD format and stores the information in the word that is 100 higher than the timer address. Figure 12–18a shows the VDT display for an ON delay timer (TON) addressed 030, and Figure 12–18b shows word 130 and the preset value of timer 030 stored in BCD format.

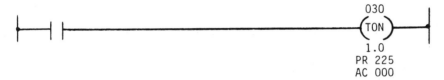

Figure 12–18a Programmed ON Delay Timer (TON)

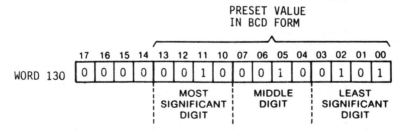

Figure 12–18b Preset Value Stored in Word 130

When using a BCD format and 12 bits, the largest number or preset time that can be stored is 999. The other four unused bits of word 130 may be used as internal storage.

As indicated earlier, when the timer is energized, the accumulated time (value) is stored in the first 12 bits of word 030 in BCD format. The VDT, however, displays the accumulated time in decimal numbers.

To better understand how the timer works, consider the circuit shown in Figure 12–19a and the equivalent program using the Allen-Bradley PLC-2 format (Figure 12–19b).

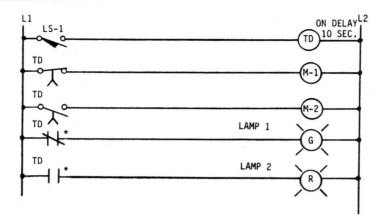

Figure 12–19a ON Delay Timing Circuit

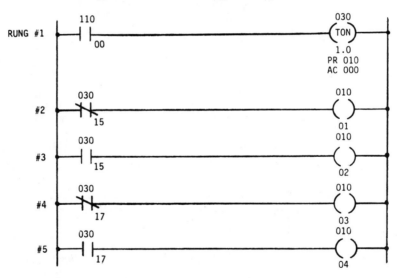

Figure 12–19b Timing Circuit Programmed Using Allen-Bradley Format

Limit switch 1 (LS-1) is programmed as a N.O. contact (EXAMINE ON) and given an input address of 11000 (word 110—bit 00). Next, an ON delay timer (TON) is programmed and addressed 030. The time base is set for seconds, 1.0, and the preset value of 10 seconds is entered (010).

In Figure 12–19a, motor 1 is controlled by a N.C. time opening contact. To program the equivalent, a N.C. (EXAMINE OFF) instruction is used and addressed 03015. Bit 15 of word 030 is a timed bit and is set to 1 (turn on) when the timer has timed out. For motor 2, which is controlled by an N.O. time closing contact, a N.O. (EXAMINE ON) instruction is programmed and also addressed 03015. Because bit 17 is turned ON, or set to 1, anytime the timer is energized, it is used for the N.O. and N.C. instantaneous contacts (Figure 12–19b).

When the processor is placed in the *run* mode, motor 1 and lamp 1 come *ON*. Both motor 1 and lamp 1 (green) are controlled by N.C. (EXAMINE OFF) contacts. As LS-1 is open, the timer is not yet energized or timed out, so bits 15 and 17 of word 030 are set to 0, or *OFF*. This makes the N.C. contacts true, so motor 1 and lamp 1, bits 01 and 03 of word 010, are *ON*. Figure 12–20 shows the bit status for words 010, 030, 110, and 130 with power *ON*, but LS-1 open.

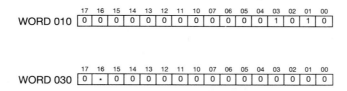

Figure 12–20 Bit status of Words 010, 030, 110, and 130

When LS-1 is closed, bit 00 of word 110 is set to 1, the timer rung is true, and the timer energizes and starts to time. With the timer energized, bit 17 of word 030 is set to 1 which makes the EXAMINE OFF (N.C.) contact (instruction) in Rung 4 go false, and the EXAMINE ON (N.O) contact (instruction) in Rung 5 go true. This turns lamp 1 *OFF* and lamp 2 *ON*.

Figure 12–21 shows the bit status for words 010, 030, 110, and 130 at the instant the timer is energized.

	17	16	15	14	13	12	11	10	07	06	05	04	03	02	01	00
WORD 010	0	0	0	0	0	0	0	0	0	0	0	0	1	0	0	1	0

	17	16	15	14	13	12	11	10	07	06	05	04	03	02	01	00
WORD 030	1	•	0	0	0	0	0	0	0	0	0	0	0	0	0	0

	17	16	15	14	13	12	11	10	07	06	05	04	03	02	01	00
WORD 110	0	0	0	0	0	0	0	0	0	0	0	0	1	0	0	1	0

	17	16	15	14	13	12	11	10	07	06	05	04	03	02	01	00
WORD 130	0	0	0	0	0	0	0	0	0	0	0	0	1	0	0	0

Figure 12–21 Bit Status of Words 010, 030, 110, and 130 the Instant the Timer Rung is True

As the timer counts in one-second intervals, the accumulated time is stored in the first 12 bits of word 030. Figure 12–22 shows the bit status for word 030 as it counts from 000, energized, to 010, timed out.

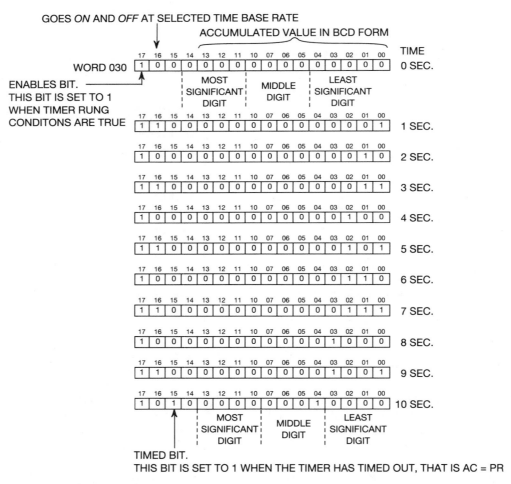

Figure 12–22 Bit Status as Accumulated Time goes from 000–010

When the timer times out (accumulated time equals the preset time), timed bit 15 of word 030 is set to 1, or *ON*. Setting bit 15 to 1 makes the EXAMINE OFF (N.C.) instruction in Rung 2 go false, turning *OFF* motor 1 (Figure 12–19b). The EXAMINE ON (N.O.) instruction in Rung 3 is now true, and motor 2, bit 02 of word 010, is turned *ON*.

Initially, when the preset time of 010 was stored in word 130, the processor compared the accumulated time in word 030 with the preset value in 130 on each scan. On the scan where the accumulated value (time) equaled the preset value (time), the processor updated bit 15 of word 030 and turned it *ON*. Anytime power is removed from the timer, the timer resets to 000.

Due to scan times, most PLCs have a timer accuracy of + or − 10m seconds. The timer accuracy decreases as scan time increases. If the accuracy is critical, the smallest time base that is available should be used (0.001 [hundredths of a second] rather than a one-second time base).

Figure 12–23a shows a circuit with a pneumatic OFF delay timer, and Figure 12–23b shows how the same circuit is programmed and addressed.

Figure 12–23a OFF Delay Timing Circuit

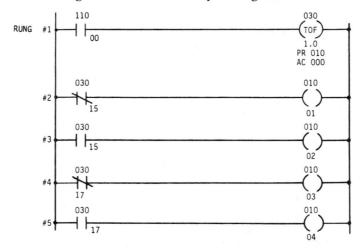

Figure 12–23b Timing Circuit Programmed Using Allen-Bradley PLC-2 Format

When the processor is turned *ON*, only Rungs 2 and 4 are true, so only motor 1 and lamp 1 are *ON*. When LS-1 (11000) closes, Rung 1 goes true, the OFF delay timer TOF-030 is energized, and bits 15 and 17 are set to 1. This makes Rungs 2 and 4 go false and Rungs 3 and 5 go true, so motor 1 and lamp 1 go *OFF*, and motor 2 and lamp 2 go *ON*. Since this is an OFF delay timer, the circuit stays this way as long as input device 11000 remains closed.

When input 11000 opens, OFF delay timer (TOF) 030 deenergizes and starts to time. At the instant the timer deenergized, bit 17 was set to 0, so Rung 4 went true and Rung 5 went false. Bit 15 continued to be set to 1, or *ON*, until the accumulated value equaled the preset value. When the values became equal, bit 15 is set to 0 and Rung 3 goes false while Rung 2 goes true.

Similar to its pneumatic relay counterpart, the OFF delay timer times out when power is removed and resets to 000 whenever power is applied.

Most PLCs also offer a timer that replaces the standard motor-driven timer. A typical motor-driven timer consists of shaft mounted cam(s) that are driven by a synchronous motor. Rotating cam(s) activate (open or close) limit or micro switches. Once power is applied, the motor turns the shaft and cam(s). The positioning of the lobes of the cam(s) and the gear reduction of the motor determine the time it takes for the motor to turn the cam far enough to activate the switches. If power is removed from the motor, the shaft stops. When power is reapplied, the motor continues turning the shaft until the switches are activated. When the timing of a device is not reset due to a loss of power, the timing is said to be **retentive**.

Retentive timers (RTO) can be programmed to replace motor-driven timers.

An RTO is shown in Figure 12–24a and a timing chart for the circuit is shown in Figure 12–24b. The retentive timer will start to time when input device 11306 is closed. If the input device is opened after three seconds, the timer accumulated value stays at 003. When input 11306 is closed again, the timer picks up the time at three seconds and continues timing. When the accumulated value equals the preset value (009), bit 15 of word 052 is set to 1, and output 01004 is turned *ON*.

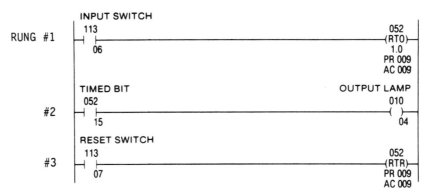

Figure 12–24a Retentive Timer Circuit
(Courtesy of Allen-Bradley)

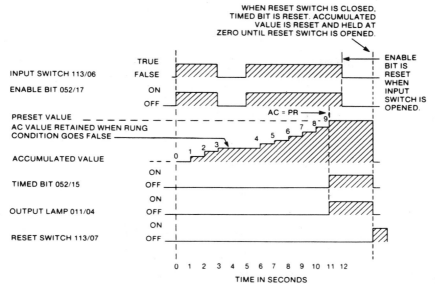

Figure 12–24b Timing Chart
(Courtesy of Allen-Bradley)

Because the retentive timer does not reset to 000 when the timer is deenergized, a reset Rung must be added. Rung 3 in Figure 12–24a illustrates how this is accomplished. The Rung consists of an N.O. input device (such as a limit switch or push button) addressed 11307 and a **retentive timer reset (RTR)**. The retentive timer reset is given the same address 052 as the RTO. When input device 11307 closes, RTR resets the accumulated value in word 052 and the timed bit 15 to 0. After resetting the RTO, input device 11307 is opened again. If input 11306 is still closed, the retentive timer will start timing again, but if input 11306 is open, RTO will remain reset to 000 until input 11306 closes. When input 11306 closes, the RTO starts to time.

When programming timers, whether they are TON, TOF, or RTO, there is no limit, except for memory size, to the number of N.O. and N.C. timed and instantaneous contacts (instructions) that can be programmed for any one timer.

Allen-Bradley PLC-5 Timers

Figure 12–25 shows the timer format for the Allen-Bradley PLC-5. The timer consists of a timing block containing the timer number (address), time base (1 second or 0.01 seconds), and the preset and accumulated times. The preset time can be programmed with any value from 0 to 32,767. If a time base of one second was assigned, 32,767 would equal 9.1 hours ($32,767 \div [60 \times 60] = 9.1$); if a time base of 0.01 (one hundredth of a second) was assigned, 32,767 would equal approximately 5.5 minutes ($32,767 \times 0.01 \div 60 = 5.46$).

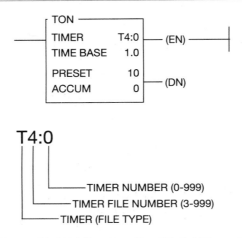

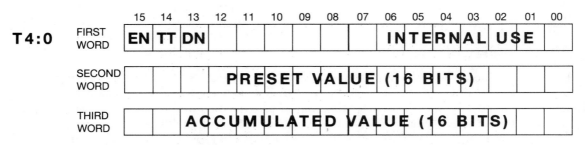

Figure 12–25 Allen-Bradley PLC-5 Timer Format

The two lines to the right of the block are the **enable (EN) bit** and **done (DN) bit** that indicate the status of the timer. The timer address (T4:0) identifies the timer file number and timer number. T4:0 indicates timer file 4, timer 0. File 4 is the default file from the data table for timers. The timer file can be assigned any number from 3–999 as long as the numbers have not be previously used or designated for other files.

NOTE: Although it is possible to use files 3 through 8 for timer files by deleting the file (or files) from the data table map, it is much easier to leave the default files alone and use files 9–999 when additional files are required.

The EN bit is set to 1 (or is true) whenever there is a logic path to the timer block. The DN bit is set to 1 (or is true) when the accumulated value equals the preset value, and the timer has timed out. Figure 12–26 shows how the information for a timer is stored. In the PLC-5, three words of memory are used for each timer programmed. The first word of memory uses the first 8 bits for internal use and uses bit 13 for the DN bit, bit 14 for the timer timing bit (TT), and bit 15 for the EN bit. The next two words store the preset and accumulated values of the timer.

Figure 12–26 PLC-5 Timer Storage Format

NOTE: There is no need to remember or memorize the timer bit numbers because the software accepts the mnemonics DN, TT, and EN. When addressing timer contacts, the timer number is entered first, then either a forward slash (/) or a period (.) is entered, followed by the timer bit. For example, T4:1/TT or T4:1.TT addresses the TT bit of timer 1, file 4.

The timer enable bit, bit 15, is set to 1, or turned *ON*, when the Rung goes true, and remains set until the Rung goes false or a reset instruction resets the timer.

NOTE: The EN bit can be used as an instantaneous contact.

The TT bit, bit 14, is set to 1, or turned *ON*, when the Rung goes true, and remains *ON* until: the Rung goes false; the DN bit is set to 1 (accumulated value = preset value); timing is completed; or a reset instruction resets the timer.

NOTE: The TT bit can be used to control a timer timing light that is only *ON* when the timer is actually timing. Figure 12–27a shows how the TT bit is used to control an indicator light, and Figure 12–27b shows the equivalent circuit using a pneumatic timer.

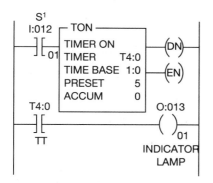

Figure 12–27a TT Bit Used to Control an Indicator Lamp

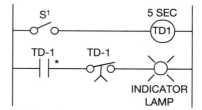

Figure 12–27b Pneumatic Timer Circuit to Control an Indicator Lamp

The DN bit, bit 13, is set to 1 when the accumulated value is equal to the preset value. The DN bit remains set to 1, or *ON*, until the Rung goes false or a reset instruction resets the timer.

NOTE: The DN bit can be used to control an output, or for other logic within a program.

Figure 12–28 shows an ON delay timer and how it is programmed to control outputs O:013/01, O:013/02, O:013/03, and O:013/04.

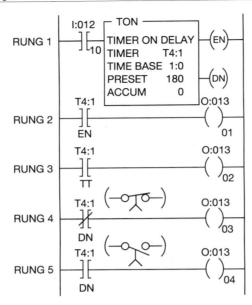

Figure 12–28 Progammed PLC-5 TON Timer

When bit I:012/10 (input device) is true, or set to 1, the timer Rung is true, and the processor starts timer T4:0 timing and sets the EN and TT bits to 1. This turns *ON* outputs O:013/01 and O:013/02 in Rungs 2 and 3. The accumulated value increases in one-second intervals. The output in Rung 4, controlled by an EXAMINE OFF instruction, is true as long as the preset is not equal to the accumulated value. The EXAMINE OFF instruction addressed with the timer DN bit acts like a normally closed-time opening contact, and does not open until the accumulated value equals the preset value. The EXAMINE ON instruction in Rung 5 with the DN bit address acts like a normally open-time closing timer contact, and does not close (or go true) until the accumulated value is equal to the preset. When the accumulated time does equal the preset time, the DN bit is set to 1, and output O:013/04 is turned *ON* and output O:013/03 is turned *OFF*. Once the timer instruction has completed timing, the TT bit is reset to 0 and the output (O:013/02) of Rung 3 is turned *OFF*.

Like a pneumatic ON delay timer, when power is removed, the timer is reset to 0. The PLC-5 timer instruction is reset when the input device (I:012/10) is opened.

Figure 12–29 shows a typical timing chart. Notice that when the Rung condition is true (*ON*), the timer will time, but if the Rung goes false (*OFF*), the timer resets to 0, as illustrated, during the first two minutes of the timing diagram.

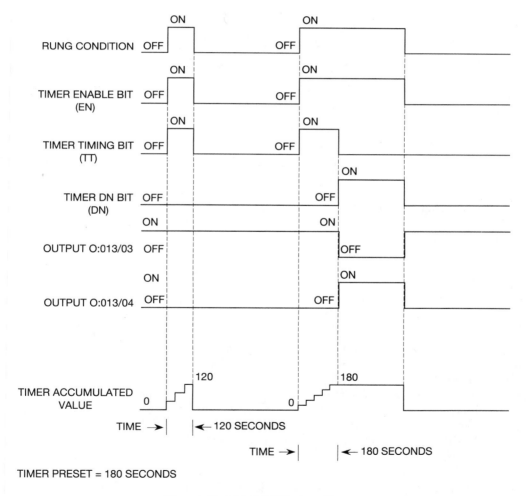

Figure 12–29 TON Timing Chart

Figure 12–30 shows how a PLC-5 OFF delay timer is programmed.

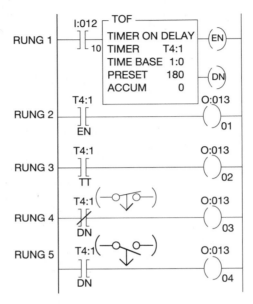

Figure 12–30 Programming a PLC-5 OFF Delay Timer (TOF)

In an OFF delay timer, when bit I:012/10 is set to 1, the DN and EN bits are also set to 1. The DN bit acts like the OFF delay contacts of a pneumatic timer, and the EXAMINE OFF (N.C.) instruction in Rung 4 goes false, or open, while the EXAMINE ON instruction (N.O.) in Rung 5 goes true. When input device I:012/10 is reset, or set to 0, the Rung is false, and the timer starts to accumulate time in one-second intervals as long as the Rung remains false. When the accumulated value equals the preset value (180) the timer stops. T4:1.TT was set to 1 while the timer was timing and output O:013/02 in Rung 3 was *ON*. When the accumulated value equaled the preset value and the timer stopped timing, the TT bit was reset to 0 and output O:013/02 was turned *OFF*. When the TT bit is reset to 0, the DN bit (bit 13) is also set to 0, and output O:013/03 in Rung 4 is turned *ON* and output O:013/04 in Rung 5 is turned *OFF*. The TOF instruction is reset by each open-to-closed transition of input device I:012/10. Figure 12–31 shows a typical timing chart for an OFF delay timer.

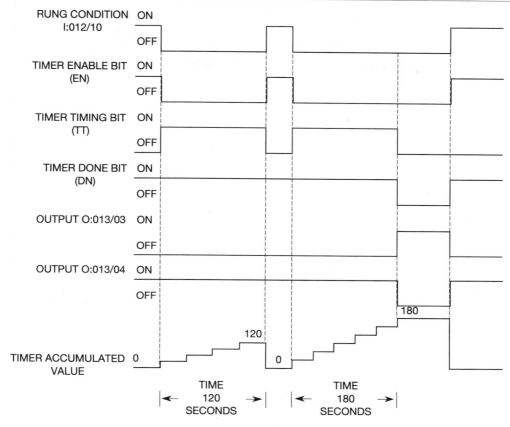

Figure 12–31 Timing Chart for a PLC-5 TOF Timer

During the first timing cycle, the timer was only *OFF* for 120 seconds. That was not long enough for the timer to time out, so the outputs controlled by the DN bit did not change. During the second timing cycle, the timer was allowed to time out and the outputs changed states.

Like the PLC-2 family, the PLC-5 family has retentive timers. The retentive timer lets the timer *START* and *STOP* without resetting the accumulated value to 0. The bits associated with the timer EN, TT and DN function the same as with the TON instruction. The RTO instruction begins timing when the Rung goes true. As long as the Rung remains true, the timer continues to time until the accumulated value reaches the preset value. If the timer Rung goes false, the timer holds the accumulated time, rather than resetting the accumulated value to 0. When the timing Rung goes true again, the count picks up from where it was, and continues to accumulate time. Once the accumulated time is equal to the preset time, the processor will set bit 13 (the DN bit) to 1. The DN bit remains *ON* (or set to 1) as long as the accumulated value is equal to the preset value. Similar to the PLC-2 RTO, the PLC-5 RTO requires a reset (RES) Rung to reset the timer. The reset instruction must be given the same address as the retentive timer it is intended to reset.

A common problem in programs that have retentive timers is that the timer is not accumulating time, even though the timer Rung is true. More often than not, the problem is a reset instruction that is true which prevents the timer from timing. Figure 12–32 shows how the RTO timer is programmed, including the reset Rung (Rung 3), and shows a typical retentive timer timing chart.

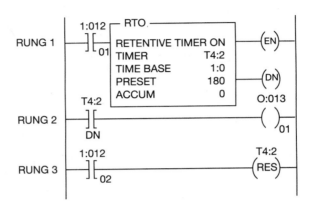

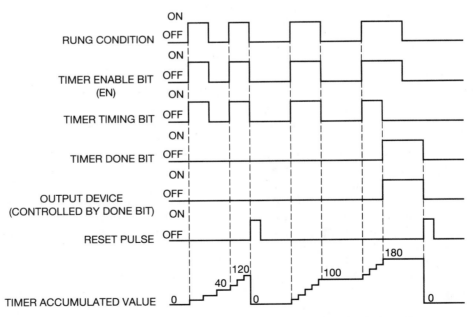

Figure 12-32 Programmed Retentive Timer (RTO) and Timing Chart

Square D Company Timers

Figure 12–33 shows a typical Square D Company timer used with their SY/MAX line of PLCs.

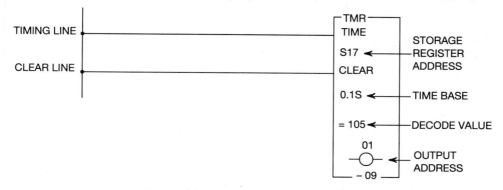

Figure 12–33 Square D Company Timer

There is a timing line, clear line, and a timer box which contains four pieces of information. The information includes the word address for a storage register that holds the timer accumulated value, the time base (0.01 seconds, 0.1 seconds, and 0.1 minute [or 6 seconds]), the decode value (preset value) which may be from 0001–9999 (counted using the binary numbering system), and an output address that is turned on when the accumulated time equals the decode (or preset) time.

Figure 12–34a shows a standard ON delay timer circuit, and Figure 12–34b shows how it is programmed.

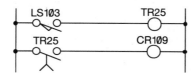

Figure 12–34a ON Delay Timer Circuit

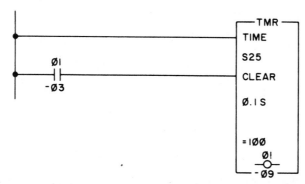

Figure 12–34b Timer Circuit Programmed Using Square D Company Format

When contact 01-03 closes, the timer times until the accumulated value in word 25 of the storage register equals the preset time of 100, and turns on output 01-09. If the input device 01-03 (LS-103) is opened before the accumulated value equals the preset or decode value, word 25 in the storage register is cleared to 0000.

Like many other PLC manufacturers, Square D does not have a dedicated OFF delay timer instruction. To program OFF delay timing action to duplicate the circuit shown in Figure 12–35, an ON delay timer is used. Before an ON delay timer is programmed to perform an OFF delay function, consider the operation in the figure, and the corresponding timing chart.

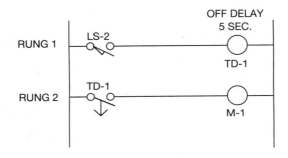

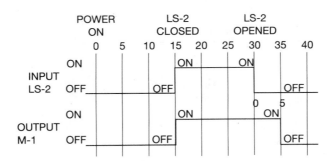

Figure 12–35 OFF Delay Timer

With LS-2 open, the OFF delay timer is not energized, even with power applied to the circuit, and M-1 is not running. When LS-2 is closed (time 15), the OFF delay timer is energized, the N.O.

time opening (T.O.) contacts in Rung 2 close, and M-1 is energized (or *ON*) as shown in the timing chart. When LS-2 is opened (time 30), the closed N.O.T.O. contacts in Rung 2 remain closed for five seconds then go open (time 35) and turn M-1 *OFF*.

To try to duplicate this circuit operation, an ON delay timer can be programmed as shown in Figure 12–36.

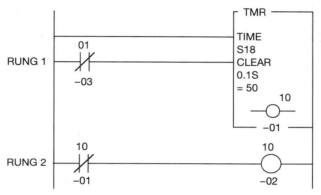

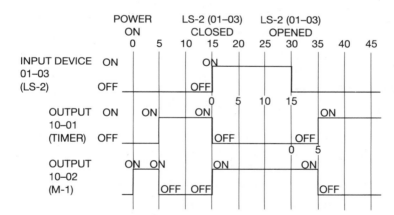

Figure 12–36 ON Delay Timer Programmed for OFF Delay Timing Action

The N.O. input contacts 01-03 (LS-2) and the N.O. time opening TD contacts are programmed N.C. (EXAMINE OFF). By using 10-01 N.C. (EXAMINE OFF) contacts in the clear line, the timer will energize and time out when power is first applied to the circuit, and open the N.C. 10-01 contacts that control output 10-02.

When the input device LS-2 (01-03) is closed, the EXAMINE OFF contacts go false (open) and the timer is reset to 0000; the N.C. timer contacts 10-01 in Rung 2 close, and output 10-02 is energized. When the input device opens again, the N.C. (EXAMINE OFF) contacts to the timer are true, and the timer starts to time. When the accumulated value in storage register S18 equals the preset value of five seconds, coil 10-01 energizes. When coil 10-01 energizes, the N.C. contacts to output 10-02 open, and 10-02 is deenergized.

The difference with the programmed OFF delay timer circuit (shown in Figure 12–35) and the ON delay timer programmed to act like an OFF delay is evidenced by the timing chart in Figure 12–36. Notice that when power is applied, 10-02 (M-1) in Rung 2 turns *ON*, even though 01-03 (LS-2) is open. With 01-03 programmed EXAMINE OFF, the timer will time when power is applied, and the 10-01 EXAMINE OFF contacts in Rung 2 will be true, or *ON*. The timer will time until the accumulated value equals the preset value (five seconds). When the accumulated value equals the preset value, coil 10-01 is energized, the EXAMINE OFF contacts in Rung 2 go false, and output 10-02 turns *OFF*. From this point on (five seconds) this circuit will duplicate the pneumatic timing circuit shown in Figure 12–35 (compare the timing charts of each circuit as LS-2 is closed [time 15 seconds] and then opened [time 30]).

The problem with this programmed circuit is that output 10-02 (M-1) is energized for five seconds when power is first applied to the circuit. Not only could this cause sequence problems, but also could represent a serious safety hazard. By reprogramming the circuit as shown in Figure 12–37, one can accurately duplicate the timing action of the circuit shown in Figure 12–35.

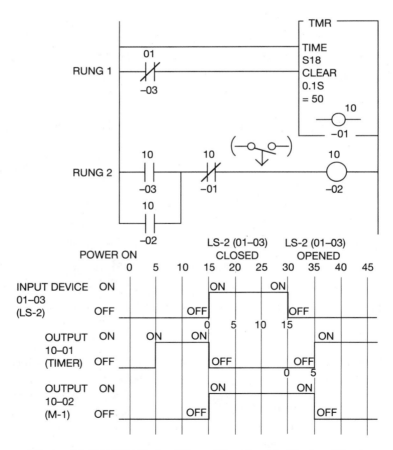

Figure 12–37 OFF Delay Timer Circuit with Timing Chart

N.C. (EXAMINE OFF) contacts 01-03 are again programmed for the N.O. input device LS-2 in the clear line of the timer (Rung 1), and an additional N.O. 01-03 contact is programmed in Rung 2. This N.O. 01-03 contact in Rung 2 prevents output 10-02 from energizing when power is applied. With the input device open, the clear line of the timer is true. The timer is activated and times until the accumulated value (S18) equals the preset value. Output 10-02 is further prevented from energizing by the now open 10-01 contacts. The timer stays activated and timed out (AC=PR) until the input device is closed.

At 15 seconds, 01-03 (LS-2) is closed which resets the timer. With the timer now deenergized, output 10-02 can energize through the now closed 01-03 contacts and the N.C. 10-01 contacts. Output 10-02 N.O. holding contacts also close. At 30 seconds, 01-03 (LS-2) opens, causing the timer to activate. Output 10-02 remains energized by its holding contacts and the N.C. 10-01 contacts from the timer. When the timer times out at 35 seconds, coil 10-01 energizes, opening the N.C. 10-01 contacts in the output circuit. Output 10-02 drops out. By not initially energizing output 10-02 when power was applied, but only energizing it when 01-03 (LS-2) closed and keeping output 10-02 energized for five seconds after 01-03 opened, exactly duplicating the OFF delay timer circuit illustrated in Figure 12–35.

To understand more fully how to program OFF delay circuits by using ON delay timers, compare the circuits and timing charts in Figures 12–38 and 12–39 using both N.O.T.O. and N.C.T.C. contacts.

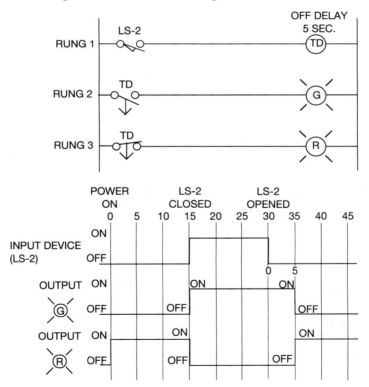

Figure 12–38 OFF Delay Timer Circuit with N.O.T.O. and N.C.T.C. Contacts

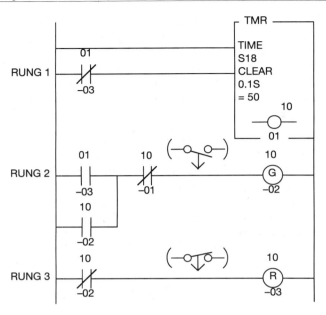

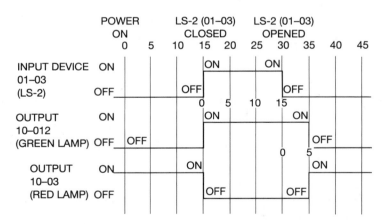

Figure 12–39 Programed OFF Delay Circuit with N.O.T.O. and N.C.T.C. Contacts

For a retentive or interruptible timer, the circuit in Figure 12–40 is used. The term *interruptible* refers to the fact that the timer retains its time, and cannot be reset to zero if power is interrupted or the logic goes false.

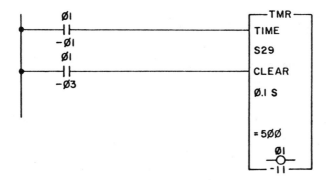

Figure 12–40 Retentive Timer

When contact 01-03 is opened, the timer value is cleared (S29=0000). To start the timer, both contacts 01-01 and 01-03 must be closed. If contact 01-01 is opened and the timer is enabled, the timer stops timing but does not clear. Reclosing contact 01-01 allows the timer to continue timing from the value at which it was interrupted, assuming contact 01-03 is still closed. When the value in register S29 is equal to the decode (preset) value of 500, the output address 01-11 will turn *ON*.

Modicon Inc. Timers

Figure 12–41 shows the timer format typical for timers programmed for the Modicon 984 family of PLCs.

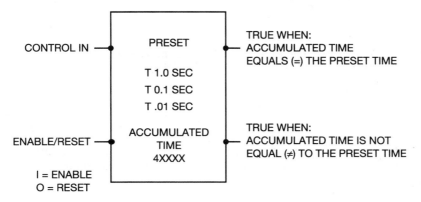

Figure 12–41 Modicon 984 Timer Format

The two lines, or nodes, to the left are **control in** and **enable/reset**. The control in line controls when the timer times, and the enable/reset line resets the accumulated value of the timer to 000, or resets the timer. The timer is enabled when both the control in and the enable/reset lines have power flow (are set to 1). The timer stops timing if the control in line goes false, but the timer is not reset until the enable/reset line goes false, or has no power flow.

The timer box holds the preset time (1–999 for some models and 1–9999 on other models). The preset value for the timer is stored in an input register (3XXXX) or in a holding register (4XXXX). The time base is selectable for seconds, tenths of a second, and hundredths of a second.

T 1.0 = seconds
T 0.1 = tenths of a second
T .01 = hundredths of a second

The timer box also displays the storage register that holds the current or accumulated time.

The two right-hand nodes, or lines, are output lines and provide power to contacts, coils, and the like. The top line provides power only when the timer's accumulated value is *equal* to the preset value; the bottom line provides power as long as the accumulated time is *less* than the preset time. The output on the bottom line only stops passing power, or is false, when the accumulated and preset values are the same (when the timer has timed out). Figure 12–42 shows an ON delay timer and how it is programmed to control outputs 00107 and 00108.

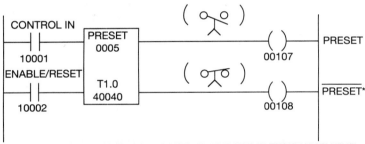

*THE LINE OVER PRESET MEANS **NOT**. THIS NODE IS **TRUE** WHEN THE ACCUMULATED VALUE IS **NOT EQUAL** TO THE PRESET

Figure 12–42 Programmed Modicon 984 Timer

If 10002 is closed, the timer is enabled, but not timing, and the value stored in register 40040 is 0. Since the value in register 40040 is 0, the accumulated value is not equal to the preset value of 00005, so output 00107 is false (*OFF*), but output 00108 is true (*ON*).

When 10001 in the control in line is closed, or goes true, the timer starts timing, and register 40040 will begin to accumulate time in one-second intervals. When the accumulated value is equal to the preset value (five), output 00107 becomes true, and output 00108 goes false. The status of the outputs remains the same until input 10002 in the enable/reset line is opened and resets the timer to zero. When the timer is reset, output 00107 is turned *OFF* and output 00108 is turned *ON*.

To duplicate the N.O.T.C. contacts of a pneumatic time delay, contacts with address 00107 should be used. For N.C.T.O. contacts, address 00108 is used.

The Gould 984, like many other PLCs, does not have a dedicated OFF delay instruction. To exactly duplicate the timing action of an OFF delay timer, additional Rungs of logic are required. Figure 12–43 shows the additional Rungs of logic required to obtain N.C.T.C. contacts and N.O.T.O. contacts. A timing chart is used to clarify the timing action.

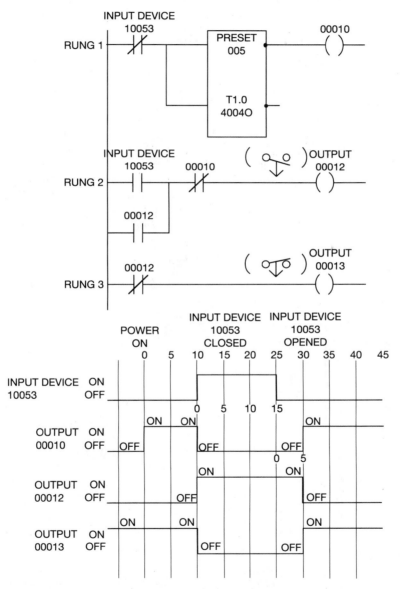

Figure 12–43 Programming a Gould 984 ON Delay Timer to Duplicate the Timing Action of an OFF Delay Timer

Notice that the enable/reset line has been tied into the control in line. This enables and turns the timer *ON* and *OFF* with just one input device (10053).

When power is first applied to the circuit, the timer is enabled and starts to time because of the EXAMINE OFF instruction 10053, which is the N.O. input device. Once the timer has timed out, output 00010 will be set to 1, or be turned ON. Output 00012 in Rung 2 cannot be energized because of the open 10053 input contacts, as well as the now open 00010 output contacts. Output 00013 in Rung 3 is energized as soon as the power is applied through the EXAMINE OFF (normally closed) contacts of output 00012. When the input device is closed, the timer Rung goes false and resets the timer. When the timer is reset, contacts 00010 in Rung 2 go back to their normally closed state. With the input device 10053 and contacts 00010 now closed, output 00012 energizes. When 00012 energizes, the EXAMINE ON contacts in Rung 2 close and act as holding contacts, while the EXAMINE OFF instruction in Rung 3 goes false and turns output 00013 *OFF*. The circuit remains in this condition until input device 10053 is opened, which activates the timer. When the accumulated time equals the preset time (value in register 40040 equals 0005), the timer will time out, making the EXAMINE OFF instruction in Rung 2 go false, which turns *OFF* output 00012, and causes output 00013 in Rung 3 to again be set to 1 (turn *ON*). This timing action exactly duplicates the timing action of an OFF delay timer. Additionally, normally open contacts (EXAMINE ON instructions) from output 00012 can be programmed for N.O.T.O. timer action, while normally closed contacts (EXAMINE OFF instructions) can be used for N.C.T.C. timer contacts.

For a retentive timer, a different input device is programmed in the control in and reset lines as indicated in Figure 12–44.

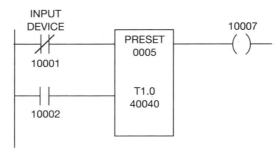

Figure 12–44 Retentive Timer

Cascading Timers

When circuit requirements demand more time than is available from a single timer, two or more timers can be programmed together as shown in Figure 12–45a. Programming two or more timers together to extend the timing range is called **cascading**.

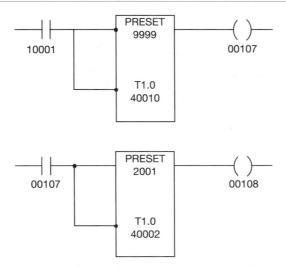

Figure 12–45a Cascading Timers

In this circuit, the first timer is controlled by input device 10001. When the device is true, the timer starts to time. When the accumulated time is equal to the preset time, output 00107 is set to 1, or *ON*. When 00107 is set to 1, the second timer is enabled and starts to time. When the second timer has timed out, output 00108 is turned *ON*. The total time to turn on output 00108 after input 10001 was closed is 12,000 seconds (9999 + 2001), or 200 minutes.

A second, and perhaps easier method to cascade timers, is to put the two timers in series as shown in Figure 12–45b. This method only works with PLCs that allow timers to be placed in series, such as the Modicon 984. If the timer is considered an output instruction, like the Allen-Bradley PLC-2 and PLC-5 families, this method does not work, and the two-Rung method shown in Figure 12–45a must be used.

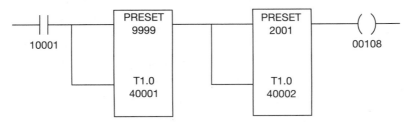

Figure 12–45b Cascading Timers in Series

Once the input device 10001 is closed, the first timer starts to time. When the first timer times out, the second timer is activated and starts to time. When the second timer's accumulated value is equal to the preset value, output 00108 will be turned *ON*. Again, the total time to turn on output 00108 after input 10001 closes, is 12,000 seconds (9999 + 2001), or 200 minutes. By cascading timers, virtually any required time can be achieved.

Chapter Summary_____

Although the format is different for different PLCs, the basic principles are the same. Preset and accumulated times are stored and compared on each processor scan. When the accumulated value equals the preset value, discrete output devices or internal outputs can be turned *ON* or *OFF*. Timers can be programmed for *ON delay*, *OFF delay*, or as *retentive* timers. The only limit to the number of timed and instantaneous contacts that can be programmed is memory size. Programmed timers offer a wider range of time settings and greater accuracy than is possible with hard-wired pneumatic timers.

REVIEW QUESTIONS_____

1. Match the standard time delay symbols.
   - a.   1.
   - b.   2.
   - c.   3.
   - d.   4.
2. The amount of time for which a timer is programmed is called the:
   - a. preset
   - b. set point
   - c. desired Time (DT)
   - d. all of the above
3. T   F   As scan time increases, so does the accuracy of any programmed timers.
4. When the timing of a device is not reset due to a loss of power, the timer is said to be:
   - a. holding
   - b. secured
   - c. retentive
   - d. continuous
5. When more time is needed than can be programmed with one timer, two or more timers can be programmed together. This programming technique is called:
   - a. stacking
   - b. cascading
   - c. doubling
   - d. synchronizing
6. When programming timers with an Allen-Bradley PLC-5, which bit will act as an instantaneous contact?
   - a. DN
   - b. TT
   - c. EN
   - d. IN

7. When the accumulated time is equal to the preset time, which bit in the Allen-Bradley PLC-5 family will be true?
   a. DN
   b. TT
   c. EN
   d. IN
8. When programming a Modicon 984, which line is true when the accumulated time is *not* equal to the preset time?
   a. preset line
   b. *not* preset line

Chapter 13 _____ Programming Counters

Objectives

After completing this chapter, you should have the knowledge to
- Write a program using up and down counters.
- Define the terms *increment* and *decrement*.

Programmed counters serve the same function as the mechanical counters used in the past. Programmed counters can count up, count down, or be combined to count up and down. Counters are similar to timers, except they do not operate on an internal clock but instead are dependent on external or program sources for counting.

Allen-Bradley Counters

Counters, like timers, compare an accumulated value to a preset value to control circuit functions. Figure 13–1 shows the 16-bit word used by the Allen-Bradley PLC-2 for an up or down counter. The PLC-2 family of PLCs has a counter range of 0–999.

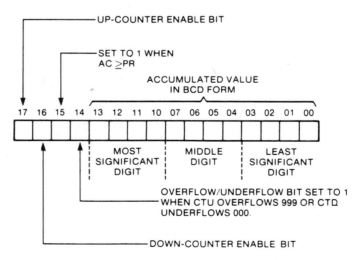

Figure 13–1 Allen-Bradley PLC-2 Counter Format

The first 12 bits (00–07 and 10–13 in the octal numbering system), are used to store the accumulated value in 3-digit BCD numbering system. Unlike a timer that stops timing when the accumulated value equals the preset value, a counter continues to count up or down. If the counter exceeds 999 or goes below 0, bit 14 will be set to 1, or go true. Bit 14 is used as an overflow (or underflow) status bit and is set to 1 when either condition, overflow (999) or underflow (000), occurs. Bit 15 is set to 1, or *ON*, any time the accumulated value is equal to (=) or greater than (>) the preset value. Bit 16 is the down counter enable bit and is set to 1 any time a down counter rung is true. Bit 17 is the up counter enable bit and is set to 1, or *ON*, when an up counter rung is true.

To activate a counter, up or down, an input device, or devices, must be used to control the logic flow (continuity) to the counter. The counter only counts up or down when there is a complete logic path to the counter. When the logic path becomes true, the counter counts up by one. It is the transition from false to true that causes the counter to **increment** (count up) or **decrement** (count down).

Input device 11105 is used in Figure 13–2 to control the count of an up counter. The CTU label stands for count up.

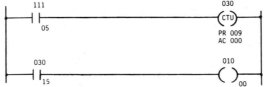

Figure 13–2 Up Counter

When input 11105 is closed, the counter *increments*, or counts up one, and the accumulated value (AC) shows 001. The counter cannot increment again until input 11105 is opened and then closed again (transition from false to true). The accumulated value increments by one with each transition of input 11105. When the accumulated value equals the preset value of 009, bit 15 of word 030 is set to 1, and output 01000 energizes. Without a reset device, the counter continues to count above the preset value of 009 with each false-to-true transition of input 11105. When the counter's accumulated value reaches 999, bit 14 is set to 1 to indicate an overflow.

To reset a counter, a **counter reset (CTR)** is added to the circuit as shown in Figure 13–3.

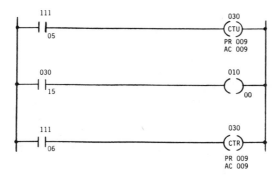

Figure 13–3 Reset Rung

Similar to the RTR (Retentive Timer Reset), the CTR is given the same word address as the counter it is resetting. When the counter reset rung is true (input 11106 is closed) the accumulated value in CTU-030 is reset to 000, and bit 15 is cleared to 0, or turned *OFF*.

Down counter (CTD) programming is illustrated in Figure 13–4.

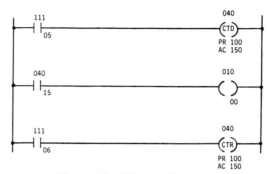

Figure 13–4 Down Counter

With the preset value of 100 and the accumulated value set to 150, the counter decrements, or counts down, from 150 by one each time input 11105 goes from false to true (*OFF* to *ON*). Since bit 15 is set to 1, or *ON*, any time the accumulated value is equal to or greater than the preset value, output 01000 is *ON*, and stays *ON* until the counter decrements to 99. In the circuit in this example, closing input 11106 and activating the counter reset (CTR-040) resets the accumulated value to 000, *not* the originally programmed accumulated value of 150. To reset the accumulated value to 150, special data manipulation techniques are used. These techniques will be covered in the next chapter.

Up and down counters can be programmed together as shown in Figure 13–5.

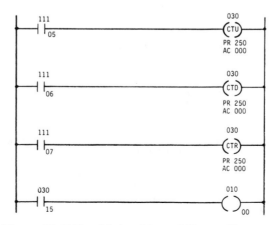

Figure 13–5 Combining Up and Down Counters

A typical application where combined up and down counters are used is in keeping count of the cars that enter and leave a parking lot or parking garage.

When a car enters the parking lot or garage, it trips (actuates) input 11105 and increments the up counter by one. Since both the up and down counters, as well as the counter reset, have the same address, the accumulated value is the same in all three. When a car leaves the parking lot or garage, it trips input 11106, and the down counter decrements or reduces the accumulated value in word 30 by one. This change is reflected in the CTU, CTD and CTR instructions. If the accumulated value reaches the preset value (AC = PR), bit 15 of word 030 is set to 1, and output 01000 is energized. Output 01000 could be a "Lot Full" sign, or any other output device that may be used to indicate all parking spaces are full.

Up, down, and combined up and down counters are retentive (hold their count) during a power failure, and are not affected by an MCR instruction.

Allen-Bradley PLC-5 Counters

The Allen-Bradley PLC-5 family offers two types of counters: **up counters (CTU)** and **down counters (CTD)**. Both counters are retentive until reset by a reset instruction. Figure 13–6 is a typical Allen-Bradley PLC-5 up counter.

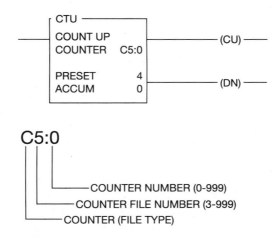

Figure 13–6 Allen-Bradley PLC-5 Counter Format

The Allen-Bradley PLC-5 up counter format is similar to the timer format. The up counter consists of a counter block that contains the up counter address, the preset value, and the accumulated count value which can be any number from –32,768 to +32,767. The counter address consists of C for counter, the file number (3–999), a colon (:), and the counter number (0–999). Each counter requires three words of memory, as shown in Figure 13–7.

NOTE: Although it is possible to use files 3 through 8 for counter files by deleting the file (or files) from the data table map, it is much easier to leave the default files alone and use files 9–999 when additional files are required.

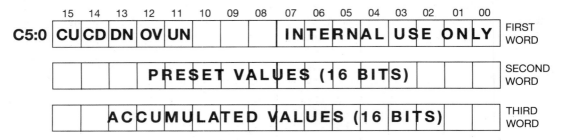

Figure 13–7 Storage Format for PLC-5 Counters

The first word stores the status bits of the counter (bits 11 through 15). The second word holds the preset values, or count, and can range from –32,768 to +32,767. The positive numbers are stored in 16-bit binary, while the negative numbers are stored in the 2s complement. The third word stores the accumulated count, and can be any number from –32,768 to +32,767. 2s complement is covered in chapter 15.

The status bits for up and down counters that are stored in the first word are as follows:

Count Up Enable Bit (CU)   The CU bit (bit 15) is set to 1, or is true, when the rung is true, and remains true as long as the up counter is enabled. The CU bit goes false when the counter is reset or the counter rung goes false. The CU bit is *only used* with up counters.

Count Down Enable Bit (CD)   The CD bit (bit 14) is set to 1, or is true, when the rung is true, and remains true as long as the down counter is enabled. The CD bit goes false when the counter is reset or the counter rung goes false. The CD bit is *only used* with down counters.

Count Up Done Bit (DN)   The DN bit (bit 13) is set to 1 when the accumulated count equals the preset count, and remains set to 1, or *ON*, as long as the accumulated value (count) is equal to or greater than the preset value.

Count Down Done Bit (DN)   The DN bit (bit 13) is *ON*, or set to 1, as long as the accumulated value is equal to or greater than the preset value. The DN bit is only reset to 0, turned *OFF*, when the accumulated count is less than the preset value.

Count Up Overflow Bit (OV)   The OV bit (bit 12) is set by the processor to 1, or *ON*, when the accumulated count exceeds the *upper limit* of (+)32,767. When this limit is reached, the count wraps around to (–)32,767, and the up counter increments from there.

Count Down Underflow Bit (UN)   The UN bit (bit 11) is set by the processor to 1, or *ON*, when the accumulated count exceeds the *lower limit* of (–)32,768. It wraps around to (+)32,767, and the CTD instruction counts down from there.

NOTE: The PLC-5 family does not use the octal numbering system for numbering its internal memory words or bits as did the PLC-2 family.

Figure 13–8 shows a CTU counter and how it is programmed to control outputs O:013/01 and O:013/02. Rung 4 is the reset rung that resets the counter's accumulated value to 0000. An output instruction is used to reset the counter. The reset (RES) command must have the same address as the counter to enable it to be reset.

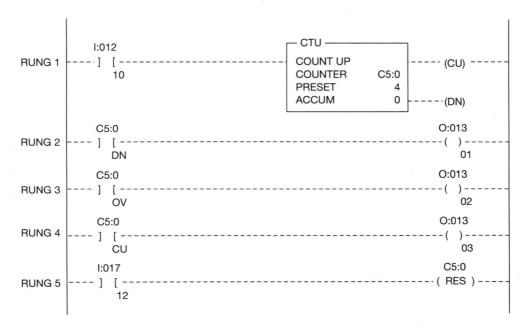

Figure 13–8 Programmed PLC-5 Up Counter (CTU)

Each time input device I:012/10 in Rung 1 makes a transition from false to true, the counter increments, or counts up by one. When the accumulated value (count) is equal to or greater than the preset count, the done (DN) bit is set to 1 by the processor, and Rung 2 becomes true, turning *ON* output O:013/01. Rung 3 is not true unless the count exceeds the counter's upper limit of (+)32,767. If the count exceeds the limit, output O:013/02 comes *ON* and remains *ON* until the counter is reset by closing input device I:017/12 in Rung 5. Bit 15, the count up bit (CU) can be programmed and used to indicated that the counter is enabled and that Rung 1 is true. The CU bit in Rung 4 is set to 1 by the processor any time that input device I:012/10 is true, thereby enabling the counter. Bit 15 is reset to 0 when Rung 1 goes false, or the timer is reset.

Figure 13–9 shows a typical counting chart for a CTU timer.

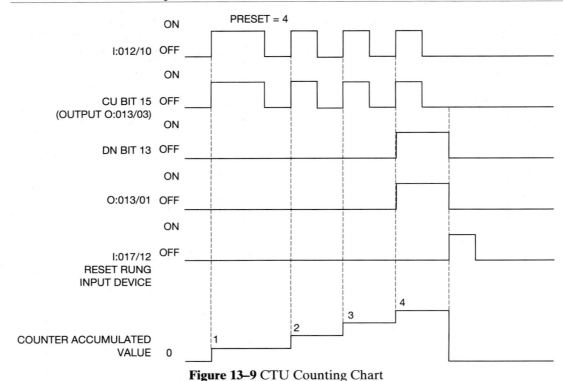

Figure 13–9 CTU Counting Chart

Figure 13–10 shows how a PLC-5 down counter (CTD) is programmed.

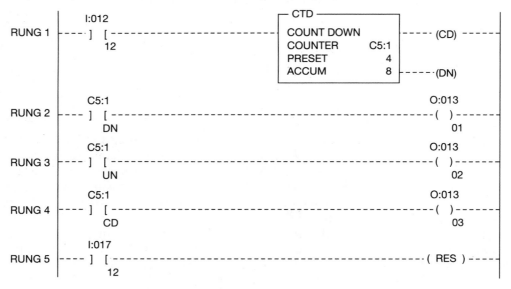

Figure 13–10 Programmed CTD Counter

The CTD counter counts down each time input device I:012/10 in the counter rung (Rung 1) goes from false to true. As long as the accumulated count is equal to or greater than the preset count, the output device (O:013/01) in Rung 2 remains energized. When the accumulated count falls below the preset count of 4 (shown in Figure 13–10), output O:013/01 is set to 0, or *OFF*. Rung 3 contains the underflow bit, which is opposite the overflow bit used with the CTU counter, and is only set to 1 when the count goes below (–)32,768. Rung 4 contains the CD bit and is *ON*, or true, any time the counter is enabled. The CD bit mirrors the status of input device I:012/12. Rung 5 is the reset rung and uses input device I:017/12 for resetting the counter. Figure 13–11 shows the counting chart for a count down timer (CTD).

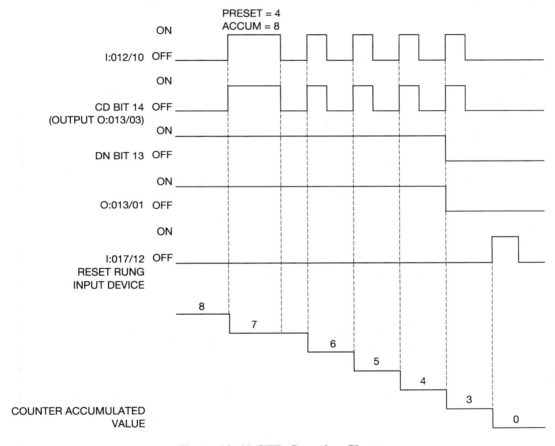

Figure 13–11 CTD Counting Chart

Figure 13–12a shows how CTU and CTD instructions can be combined, and Figure 13–12b is an example of a counting chart. When combining up and down counters, the same counter file and counter number are used for both counters as well as for the reset instruction in Rung 6.

```
         |  I:012                                        +CTU - - - - - - - - - - +           |
RUNG 1 + ---] [--------------------------------------+COUNT UP            + - (CU) - +
         |     10                                        | Counter      C5:0 |
         |                                               | Preset          4 + - (DN)
         |                                               | Accum           0 |
         |                                               + - - - - - - - - - - - +           |
         |  I:012                                        +CTD - - - - - - - - - - +           |
RUNG 2 + ---] [--------------------------------------+COUNT DOWN          + - (CD) - +
         |     11                                        | Counter      C5:0 |
         |                                               | Preset          4 + - (DN)
         |                                               | Accum           0 |
         |                                               + - - - - - - - - - - - +           |
         |  C5:0                                                         O:013 |
RUNG 3 + ---] [------------------------------------------------------( )--- +
         |    DN                                                          01 |
         |  C5:0                                                         O:-13 |
RUNG 4 + ---] [------------------------------------------------------( )--- +
         |    OV                                                          02 |
         |  C5:0                                                         O:013 |
RUNG 5 + ---] [------------------------------------------------------( )--- +
         |    UN                                                          03 |
         |  I:017                                                        C5:0 |
RUNG 6 + ---] [-----------------------------------------------------(RES) -- +
         |     12                                                            |
```

Figure 13–12a Combining CTU and CTD Instructions

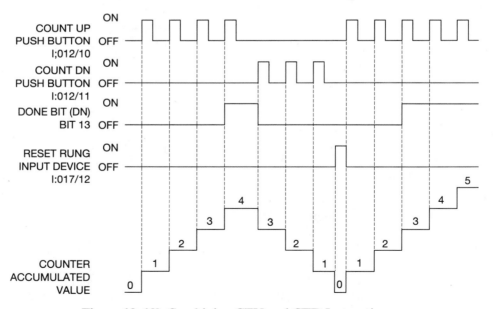

Figure 13–12b Combining CTU and CTD Instructions

Modicon 984 Counters

Figure 13–13 shows a typical Modicon 984 PLC up counter format.

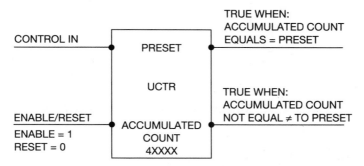

Figure 13–13 Modicon 984 Up Counter Format

The Control In line increments (increases by one) the actual count of the counter each time the line makes a false-to-true transition. The counting line (Control In line) can only increment the counter if the enable circuit is *ON*. The top line on the output side is true *only* when the accumulated count is equal to the preset value. The bottom line will be true, or *ON*, as long as the accumulated count is *not* equal to the preset count.

Figure 13–14 illustrates how an up counter (UCTR) is programmed.

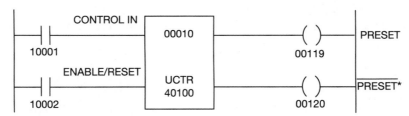

*THE LINE OVER PRESET MEANS *NOT*. THIS NODE IS *TRUE* WHEN THE ACCUMULATED VALUE IS *NOT EQUAL* TO THE PRESET

Figure 13–14 Programmed Modicon 984 Up Counter (UCTR)

When input 10002 in the enable circuit is *ON*, the counter starts from 0000 and increments the count by one each time input device 10001 in the Control In line goes from false (*OFF*) to true (*ON*). When the actual (accumulated) count stored in 40100 equals the preset count, output 00119 is energized and output 00120 is deenergized, or turned *OFF*. Any contacts labeled 00119 or 00120 in other rungs of the circuit open and/or close, based on the status of outputs 00119 and 00120. The counter is reset to 0000 when the enable circuit is opened, and is held at 0000 until the enable circuit is again closed, or true.

Figure 13–15 shows a typical Modicon 984 down counter (DCTR).

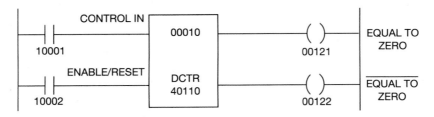

Figure 12–15 Modicon 984 Down Counter (DCTR)

Down counters count from a preset value down to 0000. When the actual count reaches 0000, output 00121 is energized (accumulated count equals zero) and output 00122 is deenergized (accumulated count is not equal to zero). When input 10002 is closed, enabling the timer, the counter counts down (decrements) from its preset value of 00010, each time the input device 10001 in the counting line goes from false (*OFF*) to true (*ON*).

When the enable circuit is opened, the accumulated value in 40110 is reset and held at the preset value of 0010 instead of 0000, like an up counter. When the enable circuit is again energized, each false-to-true transition of the counting line (or circuit) counts down from the preset value to 0000. When the actual (accumulated) count reaches 0000, output 00121 is energized and output 00122 deenergizes, or turns *OFF*.

Up and down counters can be programmed together as shown in Figures 13–16a and 13–16b to count products as they enter a conveyor line (count up) and as they leave the line (count down).

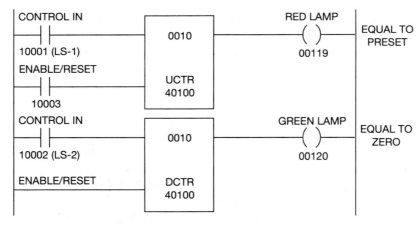

Figure 13–16A Combining Modicon 984 Up and Down Counters

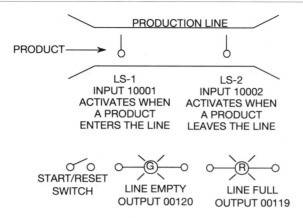

Figure 13–16b Applying Up and Down Counters

Notice that the same holding register (storage word) 40100 is used for both the up counter (output 00119) and down counter (output 00120). By sharing the same holding register, the actual value stored in 40100 of the up counter, which is programmed first, determines the actual value in 40100 of the down counter.

As a product enters the conveyor line, input 10001 is activated (false-to-true), and the actual count in 40100 increments from 0000 to 0001. The next product increments 40100 to 0002, and so on. If eight products entered the line before any were removed, the accumulated count in 40100 would be 0008. The first product to leave the line and activate 10002 decrements the actual count in 40100 (from 0008 to 0007). The next product that left the line and activated 10002 (false-to-true), would again decrease the accumulated count in 40100 from 0007 to 0006.

The indicator lamps, outputs 00120 (green) and 00119 (red) in Figure 13–16b, indicate the condition of the line. When the line is empty, the down counter has counted down to 0000, and the green lamp (output 00120) is *ON*. When the line is full, and the up counter has counted to 0010, the red lamp (output 00119) is *ON* to indicate the line is full.

Input device 10003 is used as a *START/RESET* switch. When the switch is closed, the up counter is enabled, and when the switch is opened, the up counter is reset to zero. Notice that the enable/reset line of the down counter is programmed unconditionally (no control device). This programming technique keeps the down counter enabled at all times. If the down counter was reset, it would reset to the preset value of 10, and because both counters share the same storage, or holding register, the up counter would also be reset to 10. Up and down counters are retentive during power failures.

Square D Company Counters

Figure 13–17 shows the unique approach the Square D Company uses for programming counters with their SY/MAX 100 and 300 Series PLCs.

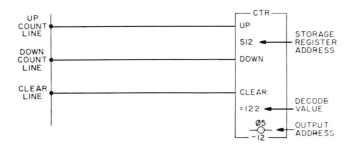

Figure 13–17 Square D Company Counter Format

This counter is a combination up and down counter. Using only the up count and clear lines, the counter becomes an up counter only. Using the down count and clear lines, it is then a down counter only. By using all three lines, it is a combination up and down counter.

Figure 13–18 illustrates how the counter is programmed to function as an up counter only.

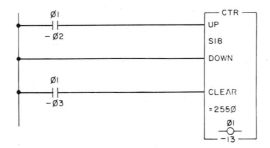

Figure 13–18 Up Counter Only

If the clear circuit input device 01-03 is closed, activating input device 01-02 in the Up Line causes a transition from *OFF* to *ON*, causing the accumulated value in storage word 18 to increment by one (from 0000 to 0001). Each subsequent transition of input device 01-02 increases the value in storage register 18 until the accumulated value equals the preset value of 2550, at which time output 01-13 is energized. If the accumulated value exceeds the preset value, output 01-13 is deenergized. With Square D counters, the output device is only *ON*, or true, when the accumulated count is equal to the preset count. If the accumulated count is less than or more than the preset count, the output is not turned *ON*.

To program a down counter, special data manipulation techniques are used. These techniques, along with an example of how to program a down counter, is covered in the next chapter.

Figure 13–19 illustrates an up and down counter by using the up, down, and clear lines.

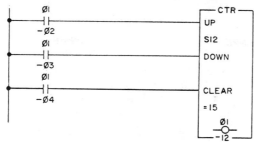

Figure 13–19 Up and Down Counter

With input device 01-04 in the clear line open, the value in storage word 12 is cleared to 0000, and output 01-12 is deenergized. When input device 01-04 in the clear line is closed, the counter is activated. Opening and closing input device 01-02 in the Up Line increments the value stored in storage register 12 by one. Open and closed transitions of input device 01-03 in the Down Line decrement (reduce) the value in storage register 12 (S12) by one. Output 01-12 does not energize until the accumulated value in S12 equals the decode (preset) value.

Up, down, and up and down counters hold the accumulated value during a power failure, or when programmed after an MCR.

Combining Timers and Counters

Timers can be combined with counters when it is necessary to extend the time of the timer beyond its normal limits. An example of combining a timer with a counter is shown in Figure 13–20.

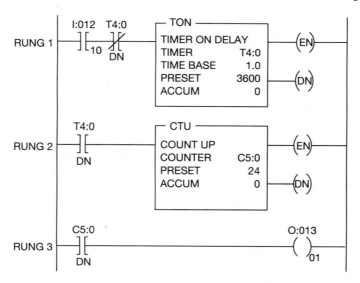

Figure 13–20 Combining a Timer with a Counter

The timer (T4:0) has a time base of 1.0 seconds and a preset value of 3600. The 3600-second preset value is equal to one hour. When input device I:012/10 is closed, the timer starts to time in 1-second increments. When the accumulated time is equal to the preset value, the DN bit (bit 13) is set to 1 and the CTU counter in Rung 2 counts, or increments, by one. (When the DN bit was set to 1, it also acted as a reset for the timer because of the EXAMINE OFF instruction in Rung 1.) The momentary action of bit 13 (false-to-true-to-false) resets the timer to 0, and the timer starts to accumulate time again. When the accumulated time on the timer has reached 3600 seconds, the timer resets itself once again and increments counter C5:0 again. The counter continues to count each time the DN bit makes a false-to-true transition (every 3600 seconds, or one hour) until the accumulated count equals the preset value of 24. When the counter has counted to 24 (24 hours), the counter DN bit is set to 1 and output O:013/01 in Rung 3 is turned *ON*.

NOTE: Remember that the length of the program effects scan time, which in turn effects timer accuracy and total time. The actual time it takes for the counter to count to 24 may be 24 hours plus or minus a few minutes.

When a timer is programmed to reset itself, as in Figure 13–20, it is referred to as a *free wheeling timer*.

Chapter Summary

Programmed counters give added flexibility and control to electrical process equipment and/or driven machinery. Similar to timers, counters store values in binary or binary coded decimal (BCD) format for the preset and accumulated counts. For up counters, the processor compares the preset and accumulated values on each scan, and updates the I/O section on the scan so that the accumulated value equals the preset value. For down counters, the processor updates the I/O section on the scan so that the accumulated value is 0 for some PLCs, or when the accumulated value is equal to or greater than the preset for other PLCs. Counters count (increment or decrement) when the count rung transition is from false to true.

REVIEW QUESTIONS

1. Define the term *increment*.
2. Define the term *decrement*.
3. What is the *preset value* or *count*?
4. What is the *accumulated value* or *count*?
5. In Figure 13–A, switch 11105 is now open. When switch 11105 is closed, counter 030 will:
   a. increment by one
   b. decrement by one
   c. not count, and the accumulated value will remain at 7

ALLEN-BRADLEY PLC-2 UP COUNTER

6. Output 02000, shown in Rung 2 of Figure 13–B, is true:
   a. only when the count is equal to the preset value
   b. when the count is equal to or greater than the preset value
   c. when the count is less than the preset value
   d. when the accumulated value reaches 999 and overflows
   e. when the count goes to 011

ALLEN-BRADLEY PLC-2 DOWN COUNTER

```
        11105                              O3O
        ─┤ ├──────────────────────────────(CTD)──
                                           PR 010
                                           AC 007
        03015                              02000
        ─┤ ├──────────────────────────────(   )──
```

7. When the accumulated count is equal to or greater than the preset count, which bit in the Allen-Bradley PLC-5 family will be true?
   a. CU
   b. CD
   c. OV
   d. DN
8. When programming a Modicon 984, which line is true when the accumulated time is not equal to the preset time?
   a. the preset line
   b. *not* the preset line
9. T   F   Up and down counters can be programmed together to count up and down.

10. The reset rung shown in Figure 13–C resets counter C5:0:
    a. automatically when the count reaches 010
    b. automatically when the count reaches 011
    c. only when the count reaches 32,767
    d. only when switch I:012/01 is closed
    e. only when switch I:012/01 is closed and then opened

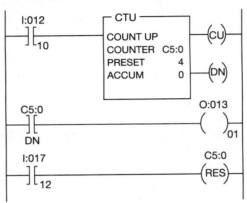

ALLEN-BRADLEY PLC-5 UP COUNTER

11. Define the term *overflow*.
12. Define the term *underflow*.
13. When an up counter accumulated value equals the preset value, the counter will:
    a. reset itself
    b. stop counting
    c. continue to count
    d. continue to count but go into an overflow condition as soon as the accumulated value exceeds the preset value

Chapter 14

Data Manipulation

Objectives

After completing this chapter, you should have the knowledge to
- Explain what data transfer is.
- Define the term *writing over*.
- Write a rung of logic that transfers data from one word to another.
- Identify the standard data compare instructions.
- Write logic that compares data to control an output.

Most PLCs now have the ability to manipulate data that is stored in memory. Data manipulation can be placed in two broad categories: data transfer and data compare.

Data Transfer

Data transfer consists of moving or transferring numeric information stored in one memory word location to another word in a different location. Words in the user memory portion of the processor may be referred to as **data table words**, and **holding register** and/or **storage register words**, depending on the PLC.

Figures 14–1a and 14–1b illustrate the concept of moving numerical data from one word location to another word location. Figure 14–1a shows that numeric (binary) data is stored in word 18, and that no information is currently stored in word 31.

16	15	14	13	12	11	10	9	8	7	6	5	4	3	2	1
1	1	0	1	0	0	1	1	0	1	0	1	1	1	0	1

WORD 18

16	15	14	13	12	11	10	9	8	7	6	5	4	3	2	1
0	0	0	0	0	0	0	0	0	0	0	0	0	0	0	0

WORD 31

Figure 14–1a Numeric Data Stored in Word 18

16	15	14	13	12	11	10	9	8	7	6	5	4	3	2	1
1	1	0	1	0	0	1	1	0	1	0	1	1	1	0	1

WORD 18

16	15	14	13	12	11	10	9	8	7	6	5	4	3	2	1
1	1	0	1	0	0	1	1	0	1	0	1	1	1	0	1

WORD 31

Figure 14–1b Data Transferred From Word 18 Into Word 31

After the data transfer (Figure 14–1b), word 31 now holds the exact or duplicate information that is in word 18. If word 31 had information already stored, rather than all 0s, the information would have been replaced. When new data replaces existing data in a word after a transfer, it is referred to as **writing over** the existing data.

Chapter 13 of this text discussed the Allen-Bradley PLC-2 down counter, and indicated that by using special programming techniques, the counter's accumulated value could be reset to some value other than 000. This is accomplished by using two instructions: *GET* [G] and *PUT* (Put).

GET instructions tell the processor to "get" a value stored in some word. The *GET* instruction shown in Figure 14–2a tells the processor to get the value 150 that is stored in word 020.

The *PUT* instruction tells the processor where to "put" the information it obtained from the *GET* instruction.

Figure 14–2a Allen-Bradley *GET* Instruction

The *GET* and *PUT* instructions in Figure 14–2b tell the processor to *get* the numeric value 150 stored in word 020 and *put* it into word 070. *PUT* instructions must be preceded by a *GET* instruction(s).

Figure 14–2b *GET* and *PUT* Instruction

Figure 14–3 shows an Allen-Bradley down counter circuit with a rung containing *GET* and *PUT* instructions for resetting the accumulated value of down counter 040 to 150.

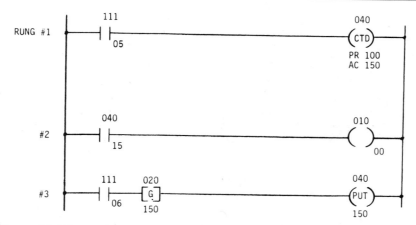

Figure 14–3 Resetting a Down Counter Using *GET* and *PUT* Instructions

After programming input device 11106, the *GET* key [G] is pressed, and can now be addressed with an unused word in memory. For this example, word 20 is used. The desired numeric value of 150 is entered and stored in word 20. The *PUT* key (Put) is now pressed, and the address 040 entered because that is where the accumulated value for down counter 040 is stored.

NOTE: Remember that in the PLC-2 family, accumulated values are always stored in the word address of the Allen-Bradley timer or counter, whereas the preset values are automatically stored 100 words higher.

Rung 3 says: if input device 11106 is closed, *GET* the numerical value stored in word 20 (150) and *PUT* it into word 40.

NOTE: In Allen-Bradley PLC-2 down counters, bit 15 is set to 1, or *ON*, any time the accumulated value is greater than or equal to the preset value. Output 01000 in Figure 14–3 is *ON* until the time counts down to 099 (less than and not equal to 100).

When the counter has counted down from 150 to 099 and deenergized output 01000, the counter's accumulated value can be reset to 150, rather than to 000, by momentarily closing input 11106. If input device 11106 is left closed, the *PUT* instruction holds the accumulated value at 150 and the accumulated count is not able to change value, even if the counter is being activated by input 11105 in Rung 1. To eliminate the possibility of holding, or freezing, the accumulated value, the circuit shown in Figure 14–4 could be programmed to ensure that the *GET* and *PUT* combination is true for only one processor scan.

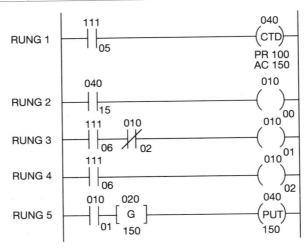

Figure 14–4 Circuit to Limit *GET* and *PUT* Combination to One Scan (One Shot Circuit).

When input device 11106 in Rungs 3 and 4 is closed, output 01001 (Rung 3) and output 01002 (Rung 4) are turned *ON*. On the next scan, the 01001 EXAMINE ON instruction in Rung 5 is set to 1, and the *GET* and *PUT* instructions will be true. A value of 150 is placed in the accumulated value of CTD 40. On the next scan, the 01002 EXAMINE OFF instruction in Rung 3 is set to 0 and held at 0 until input device 11106 makes a false-to-true transition. With 01001 set to *OFF*, the *GET* and *PUT* instructions in Rung 5 are false, and was true for only one program scan. This programming technique allows accumulated values to be placed in PLC-2 counters and timers without worrying about the accumulated value being frozen or locked in place if the initiating device is left in the closed position.

A circuit that limits logic to only one scan is often referred to as a **one shot circuit**. As discussed in Chapter 11, the Allen-Bradley PLC-5 has an instruction called one shot (*ONS*). This instruction automatically limits the output to one scan and eliminates the necessity of programming extra rungs to obtain the desired results.

With the *GET* and *PUT* instructions, data in any storage word can be transferred to other words for accumulated *or* preset values. One example is a counter that counts boards at a sawmill. When the mill is producing 2X4s, it wants 400 in a stack. However, when the mill is producing 2X6s, only 250 boards are needed for a full stack. Figures 14–5a and 14–5b show how to change the preset value of an up counter for each different lumber size by using push buttons.

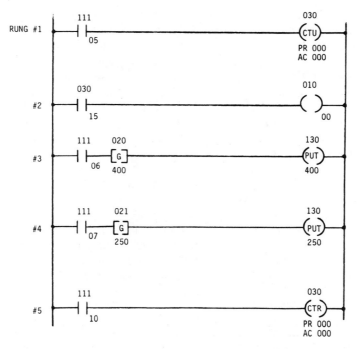

Figure 14–5a Changing Preset Values with *GET* and *PUT* Instructions

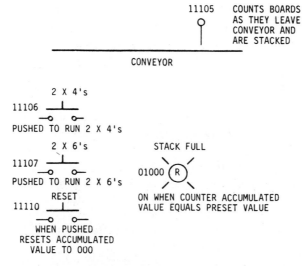

Figure 14–5b Push Buttons Used to Change Preset Values

Up counter 030 is initially programmed with no preset (000). The preset is determined by which-ever push button is depressed. If 2X4s are to be counted, push button 11106 is pushed, enabling the *GET* and *PUT* statements in Rung 3. When this rung is enabled, or true, it tells the processor to *GET* the value 400 stored in word 020 and *PUT* it into word 130. This causes counter 030 to be preset to 400. A push button is used so the rung will go false (open) after the preset value has been set. Holding the button down and keeping Rung 3 true, holds the value at 400, and transitions of input device 11105 do not increment the counter. After 400 boards have been counted (PR = AC), bit 15 of word 030 in the up counter will be set to 1, and the "Stack Full" light, 01000, comes *ON*. After the stack has been moved, counter reset push button 11110 is pushed to clear the accumulated value back to 000.

To change the preset value of up counter 030 from 400 to 250, the 2X6s' push button (11107) is depressed.

Square D Company uses a *LET* instruction with their SY/MAX PLCs to achieve the same results. To use a *LET* instruction, a storage word is addressed, and the desired value is entered. Chapter 13 skipped over the Square D down counter and indicated that it took a special pro-gramming technique. Figure 14–6 shows a down counter and a *LET* rung programmed for pre-setting the accumulated value.

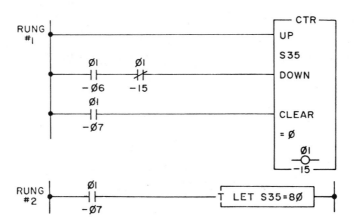

Figure 14–6 Presetting the Accumulated Value of a
Square D Company Down Counter

The *LET* rung is programmed with the same input device, 01-07, in the Clear Line. With input 01-07 open, the accumulated value of the counter, stored in word 35, is cleared to 0000, and out-put address 01-15 is *OFF*. When input 01-07 is closed, the counter is enabled, and the *LET* rung also goes true. The *LET* statement says: *LET* storage word 35 = 80. When Rung 2 is true, a value of 80 is entered or put into word 35. This sets the accumulated value of the down counter at 80.

NOTE: The *LET* instruction shown is transitional, as indicated by the T in the left margin of the *LET* box. A transitional *LET* instruction operates only once on each transition (open-to-closed) of input device 01-07. This allows the accumulated value to be set, but not held, at 80. This instruction is similar to the Allen-Bradley PLC-5 one shot instruction.

Each time 01-06 has a transition from false-to-true, the counter decrements by one. After 80 transitions of input 01-06, the accumulated value equals zero (AC = PR), and output 01-15 turns *ON*. When output 01-15 energizes, it opens the N.C. contact in the down line which disables the counter. Any additional transition of contacts 01-06 is ignored. To turn *OFF* output 01-15 and reset the counter, input device 01-07 is opened and then closed again. With the accumulated value again set to 80 by the *LET* rung, input device 01-06 again decrements the counter by one with each transition from *OFF* to *ON* or false to true.

A regular nontransitional *LET* instruction is shown in Figure 14–7.

Figure 14–7 Nontransitional *LET* Statement

When contacts 01-04 and 01-08 are closed, the value 1535 is preset into storage register S42. The *LET* is performed with each scan of the processor memory as long as 01-04 and 01-08 remain closed.

Data transfer from one storage word to another is illustrated in Figure 14–8.

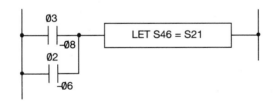

Figure 14–8 Transferring Data from One Storage Word to Another

If either contact 03-08 or 02-06 is closed, the value in storage register S21 will be transferred into storage register S46.

Allen-Bradley PLC-5 Data Transfer Instructions

The PLC-5 family use a **move (MOV)** instruction for moving data from one word to another. *MOV* is an output instruction that copies a value from one word (source address) to another word (destination address). When the rung that holds the *MOV* instruction is true, the instruction moves data from the source address into the destination address on each processor scan. The PLC-5 *MOV* instruction combines the *GET* and *PUT* instructions used on the PLC-2 into one instruction. Figure 14–9 show the *MOV* format.

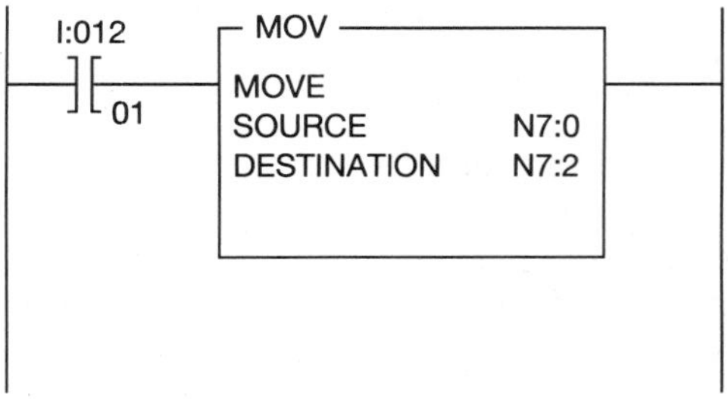

```
    I:012        ┌─ MOV ──────────────────────┐
  ──┤ ├──────────┤  MOVE                        ├──
       01        │  SOURCE           N7:0        │
                 │  DESTINATION      N7:2        │
                 └──────────────────────────────┘
```

Figure 14–9 PLC-5 *MOV* Format

When input device I:012/01 is closed and the rung is true, the *MOV* instruction reads the data from the source address and copies it to the destination address. In this case, the source is integer file N, file number 7, and word 0. The destination is integer file N, file 7, word 2. As long as the rung stays true, the values found in word 0 of N7 is transferred (copied) into word 2 of N7 on each program scan. The integer file (as discussed in Chapter 5) is used in the PLC-5 for storing whole numbers.

Another PLC-5 data manipulation instruction is the **masked move (MVM)** instruction. The *MVM* is an output instruction that copies a value from a source address to a destination address, but in addition, allows portions of the data to be masked, or blocked from being copied. The format for the *MVM* instruction is shown in Figure 14–10.

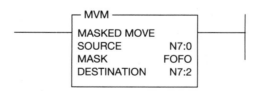

Figure 14–10 Masked Move (MVM) Instruction Format

To program an *MVM* instruction, a source address and a destination address are required, just as in the *MOV* instruction. The additional requirement for the *MVM* instruction is the mask data. For each bit of the destination word that is to be masked, or not copied, a 0 is used. If, on the other hand, it is desired that the data from the source word be written into specific bits of the destination word, a 1 is placed in that bit location. Figure 14–11 clarifies the operation of the *MVM* instruction.

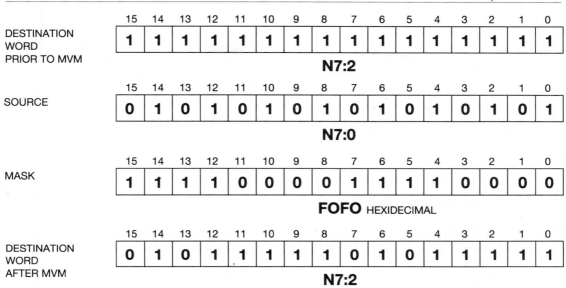

Figure 14–11 Mask Bits Used to Block Transfer of Data from Source Address
Into Destination Address

For each location in the destination word that you want to be overwritten by the data from the source word, a 1 is used. In Figure 14–11, only bits 4 through 7, and bits 12 through 15, are set to 1 in the mask, so only these bits of the destination word will have data transferred in from the source word. Those bits in the destination word that had 0s in the mask (bits 0 through 3 and bits 8 through 11) remain unchanged when the *MVM* instruction is true.

The bit status for the mask is entered by addressing a word and file that has the desired bit order that is wanted; for example, B100:0 (Binary file, file 100, word 0). The value can also be entered into the instruction using the hexadecimal format. The mask bit pattern shown in Figure 14–11 is F0F0 in hexadecimal.

The Modicon 984 PLC uses a **data transfer**, or **DX**, instruction to move data from one word to another. The *DX* instruction can also be used to make file-to-word moves and word-to-file moves, which will be covered in more detail in Chapter 16.

The *GET* and *PUT*, *LET*, *MOV*, *MVM*, and *DX* instructions (as well as other designated instructions by different PLC manufacturers) are all *data transfer instructions*, and their objective is to move numerical information from one word into another. To illustrate this concept, the data transfer for entering accumulated and preset values into counters was used. This is by no means the only application for a data transfer instruction. When using a given PLC, and becoming more familiar with its operation and capabilities, data transfer becomes a powerful tool that has many applications.

Data Compare

Data compare opens a new realm of programming possibilities, and demonstrates why PLCs are rapidly replacing most, if not all, hard-wired control systems.

Data compare instructions, as the name implies, compare the data stored in two or more words and make decisions based on the program instructions.

Numeric values in two words of memory can be compared for *less than* (<), *equal to* (=), *greater than* (>), *less than or equal to* (≤), *greater than or equal to* (≥), and *not equal to* (≠), depending on the PLC.

Data compare concepts were previously used when timers and counters were discussed.

The ON delay timer turns *ON* an output when the accumulated value equals the preset value (AC = PR). What happens is the accumulated numeric data in one memory word is compared to the preset value in another word on each scan of the processor, and when the accumulated value equals the preset value (AC = PR), the output is turned *ON*. Additional programming instructions can compare memory words and turn *ON* outputs when the values are less than (<), equal to (=), greater than (>), and so on.

To graphically demonstrate how data compare instructions can be used, consider the hard-wired circuit in Figure 14–12. This circuit uses three pneumatic time delay relays to start up a 4-motor conveyor system in inverse order (4–3–2–1).

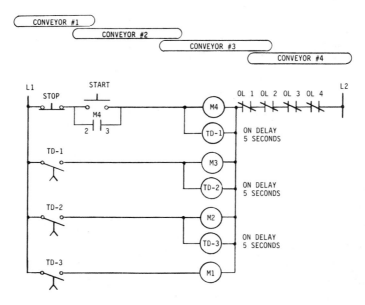

Figure 14–12 Hard-Wired Conveyor System

The same circuit can be programmed on a Square D Company PLC using only one internal timer and two data compare statements, as shown in Figure 14–13.

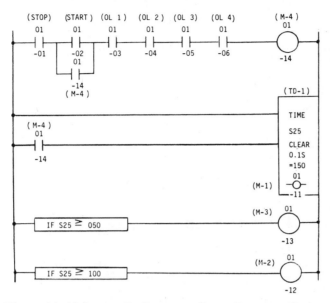

Figure 14–13 Square D Company Data Compare Format

Assume that *STOP* button 01-01 and overloads 01-03, 01-04, 01-05, and 01-06 are closed. When the *START* button (01-02) is pushed, M-4 contacts 01-14 energize, and holding contacts 01-14 close and hold the circuit in. M-4 contacts 01-14 close and enable the timer. The timer has been preset to 15 seconds (0.1 second time base × preset 150 = 15.0 seconds). The accumulated time is stored in word 25 of the storage register (S), and output 01-11 (M-1) energizes when the accumulated value in S25 equals the preset value of 15 seconds.

The *IF* instruction preceding output 01-13 (M-3) says: if the numeric value stored in word S25 becomes equal to or greater than (≥) 050 (5 seconds), output 01-13 (M-3) should be turned *ON*. When the accumulated value in word S25 reaches 050, it is equal to (=) 050, and output 01-13 (M-3) turns *ON*. As the accumulated value in S25 goes to 051, the value is now greater than (>) 050, so M-3 remains energized, or *ON*.

As the timer continues to time, the *IF* instruction preceding output 01-12 (M-2) is true when the accumulated value in S25 is equal to or greater than (≥) 100 (10 seconds), output 01-12 turns *ON*.

M-1 turns *ON* when the accumulated value in S25 equals the preset value of 150 (15 seconds).

Not only does this technique give added flexibility for timers, but also saves memory. Programming the circuit in Figure 14–12 with three on delay timers would have used 18 words of user memory, because each timer requires six words of memory. *IF* and *LET* statements only use three words each, so by programming the circuit as shown in Figure 14–13, only 12 words of memory were used for the timing function. Six words are required for the timer and three words for each of the two *IF* statements (6 + 6 = 12).

The data comparison symbols used with an *IF* instruction in the Square D SY/MAX PLCs are: equal to (=); not equal to (≠); greater than or equal to (≥); and less than (<).

Figure 14–14 is a circuit that illustrates how each instruction operates.

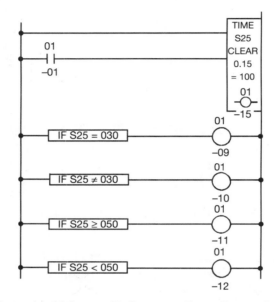

Figure 14–14 Square D Company Data Comparisons

The timer is set for 10 seconds (0.1 seconds × 100).

When power is applied, but before 01-01 is closed to activate the timer, outputs 01-10 and 01-12 are energized. The *IF* statement preceding output 01-10 is true if storage register word 25 is (≠) to 030. With 01-01 open, S25 is reset to 000 and is (≠) to 030. Output 01-12 is energized because the *IF* statement is true any time the value in S25 is (<) 050.

When input 01-01 closes, the timer is enabled and starts to time. At time 030, the *IF* statement preceding 01-09 goes true because the accumulated value in S25 is (=) 030, and output 01-09 is energized. This is only true when the accumulated value in S25 is 030. When it advances to 031, the *IF* statement goes false and 01-09 goes *OFF*. Output 01-10, which was *ON* because of the (≠) *IF* statement, now goes *OFF* for one second because the *IF* statement was false when the value in S25 was (=) 030.

At 050, 5 seconds, output 01-11 turn *ON* because the *IF* statement preceding it goes true when S25 is (≥) 050. The rung is true when the accumulated value in S25 is (=) 050, and remains true as long as the accumulated value is 050 or greater. Output 01-11 remains *ON* until the timer is cleared and the accumulated value in S25 is reset to 000.

Output 01-12, that was *ON*, now goes *OFF* at 050 because the *IF* statement that precedes it is only true when the value of S25 is (<) 050.

Output 01-15, the timer output, comes *ON* at 100 when the accumulated value equals the preset value. The time chart in Figure 14–15 illustrates the *ON* and *OFF* states of the outputs in relation to time and the *IF* instructions.

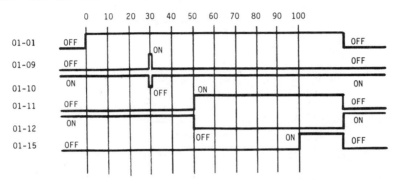

Figure 14–15 Time Chart for Data Comparisons

Data comparisons can also be made with the value in one storage register to a value in another storage register (Figure 14–16).

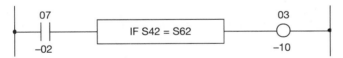

Figure 14–16 Comparing Data in Storage Registers

If input device 07-02 is closed, and if the value in storage word 42 is (=) the value in storage word 62, output 03-10 is energized.

In the timer circuit shown in Figure 14–13, a number in a storage register was compared to a constant. Figure 14–17 is another example of this concept. A value in a storage register can also be compared to the results of a math operation. Math and arithmetic operations are covered in detail in Chapter 15.

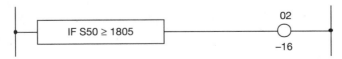

Figure 14–17 Comparing Data in a Storage Register to a Constant

Data comparison instructions can be programmed in series, parallel, or series parallel on the same rung for added circuit control. Figures 14–18a and 14–18b show *IF* statements programmed in series and parallel.

Figure 14–18a Data Comparisons Programmed in Series

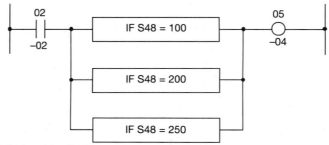

Figure 14–18b Data Comparisons Programmed in Parallel

The basic Allen-Bradley PLC-2 data comparison instructions are less than ($<$) and equal to ($=$). But by combining instructions, data comparisons can also be made for greater than ($>$), less than or equal to ($\leq$), greater than or equal to ($\geq$), and not equal to ($\neq$).

A *GET* [G] instruction must be programmed prior to any data comparison instruction.

An equal to ($=$) data comparison example is shown in Figure 14–19.

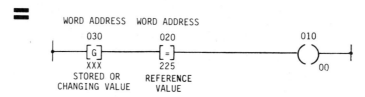

Figure 14–19 Allen-Bradley PLC-2 Equal To Comparison

First a *GET* instruction is programmed and given the address of the word that stores the value to be compared. The value, represented by XXX, could be a stored value in memory or a changing accumulated value of a timer or counter. Next, the data comparison instruction is entered and given the address of the word in which the reference or comparison value is to be stored. In the example (Figure 14–15), output 01000 turns *ON* when the value in word 030 is equal to the reference value 225 in word 020. Output 01000 deenergizes if the value in 030 changes to 226 because the value is no longer equal to the reference value of 225.

For a less than ($<$) data comparison, the circuit is programmed as shown in Figure 14–20.

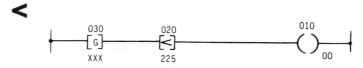

Figure 14–20 Allen-Bradley PLC-2 Less Than Comparison

Output 01000 is *ON* as long as the value of word 030 is (<) the reference value 225 of word 020, and goes *OFF* when the value is (≥) 225.

To program a (>) data comparison, reverse logic is used. As stated earlier, there are only two basic data comparison instructions ([=] and [<]), when programming the Allen-Bradley PLC-2 family. The (<) instruction is used for a (>) comparison and addressed with the word that is to be compared. The [G] instruction is addressed with the word that is to store the reference value. A (>) circuit is shown in Figure 14–21.

Figure 14–21 Allen-Bradley PLC-2 Greater Than Comparison

For a (>) comparison, the value in word 030 (the less value) is compared to the reference value of word 020 (the *GET* value), and the rung is only true when the less value is greater than the *GET* value of 225.

For a (≤) comparison, the rung is programmed with one *GET* instruction followed by less than (<) and equal to (=) instructions in parallel (Figure 14–22).

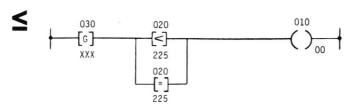

Figure 14–22 Allen-Bradley PLC-2 Less Than or Equal To Instructions

Output 01000 is *ON* as long as the value in word 030 is (<) 225. When the value in word 030 reaches 225, the (=) instruction is true, but goes false at 226.

NOTE: Both the (<) and (=) instructions can use the same word address and reference number.

Figures 14–23 and 14–24 illustrate greater than or equal to (≥) and not equal to (≠) circuits.

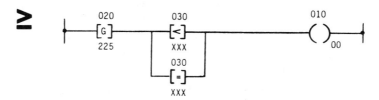

Figure 14-23 Allen-Bradley PLC-2 Greater Than or Equal To Instruction

Output 01000 comes on when the value in word 030 equals the reference value 225 in word 020, and stays *ON* if the value goes to 226 because it is now greater than 225.

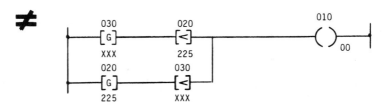

Figure 14-24 Allen-Bradley PLC-2 Not Equal To Instruction

Greater than and less than statements are programmed in parallel so that output 01000 can energize as long as the value in 030 is (<) 225 or if the value in 030 goes to 226, which is greater than 225. Output 01000 will deenergize when the value of 030 is equal to the reference value (225) in word 020.

Allen-Bradley PLC-5 Data Compare Instructions

The Allen-Bradley PLC-5 family of programmable controllers has a set of data compare instructions that are different from those used with the PLC-2 family, but the principles are the same. The PLC-5 instructions include *equal* (EQU), *greater than or equal* (GEQ), *greater than* (GRT), *less than or equal* (LEQ), and *less than* (LES). Figure 14-25 shows how the EQU instruction is programmed.

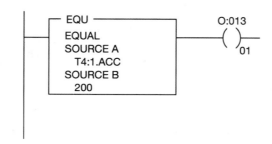

Figure 14-25 Allen-Bradley PLC-5 Equal To Instruction

The EQU instruction is true, and turns *ON* output O:013/01 when the value in Source A is equal to the value in Source B. Source A and B can be either numeric values or addresses that contain values. The value in Source A is the value of address T4:1.ACC (timer file 4, timer 1, accumulated value), whereas the value in Source B is the numeric value 200. To use either the accumulated or preset values of timers and counters, a period is entered after the timer number, followed by ACC or PRE.

Figure 14–26 illustrates how the GEQ instruction operates.

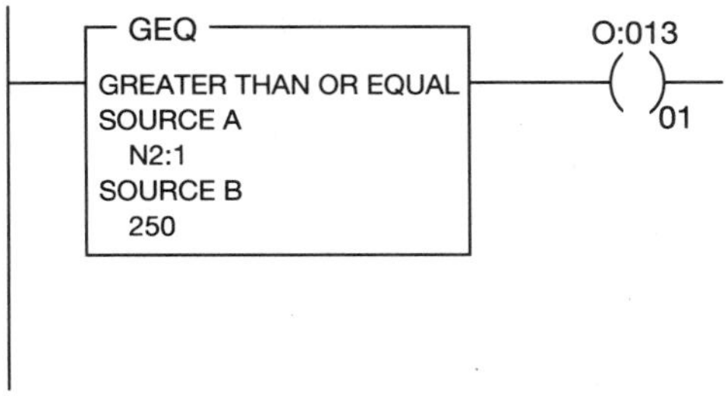

Figure 14–26 Allen-Bradley PLC-5 Greater Than or Equal To Instruction

This instruction becomes true and turns output O:013/01 *ON* when the value in Source A is greater than or equal to the value in Source B. Again, the value that is in Source A or B can be numeric values or addresses that contain values. In this illustration, the value in Source A is the value stored in integer File 7, word 1. The value in Source B is the numeric value of 250.

Another instruction is GRT (greater than) and is programmed as shown in Figure 14–27. This instruction is true when the value in Source A is greater than the value in Source B.

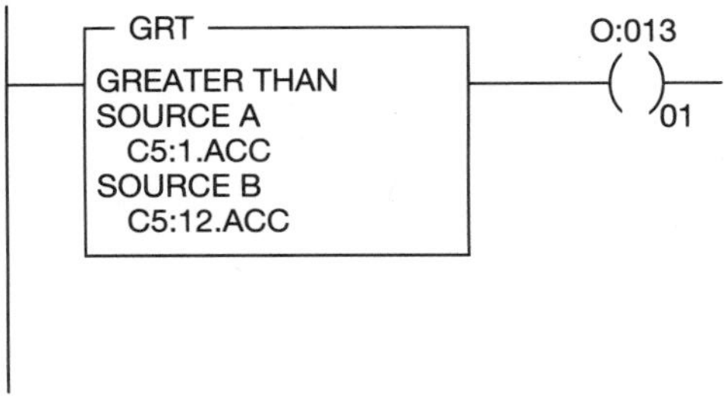

Figure 14–27 Allen-Bradley Greater Than Instruction

The instruction is true as long as the value in Source A is greater than the value in Source B. In Figure 14–27, output O:013/01 is turned *ON* anytime that the accumulated value of counter C5:1 is greater than the accumulated value in counter C5:12. As in the earlier example, the preset and accumulated values of timers and counters can be referenced by typing a period followed by either ACC or PRE after the timer or counter address. In Figure 14–27, Source A is the accumulated value of Counter 1, in counter file 5 (C5:1.ACC). Source B is the ACC value of timer 12, in counter file 5 (C5:12.ACC).

The less than or equal instruction (LEQ) is programmed as shown in Figure 14–28. This instruction is true whenever the value is Source A is less than or equal to the value stored in Source B.

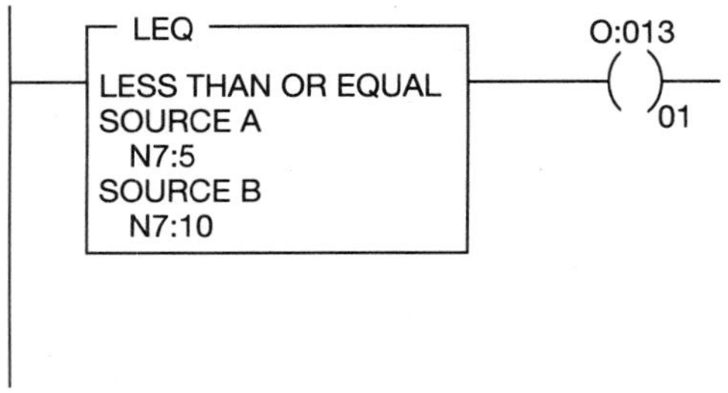

Figure 14–28 Allen-Bradley PLC-5 Less Than or Equal To Instruction

The LEQ instruction is true as long as the value in N7:5 is less than or equal to the value in N7:10. When the value in Source A is less than or equal to the value in Source B, output O:013/01 is turned *ON* by the processor.

The LES (less than) instruction is logically true when the value in Source A is less than the value in Source B. Figure 14–29 shows a LES instruction.

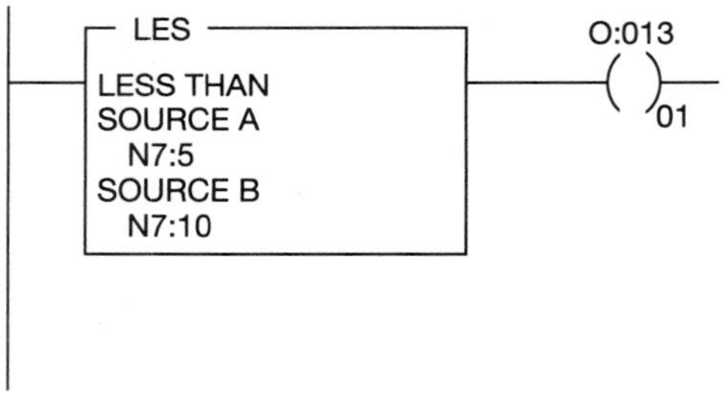

Figure 14–29 Allen-Bradley Programmed PLC-5 Less Than Instruction

In Figure 14–29, output O:013/01 is only *ON* when the value in N7:5 (Source A) is less than the value in N7:10 (Source B). When the value in Source A is larger than the value in Source B, the instruction is not logically true, and output O:013/01 is set to 0, or *OFF*.

Another PLC-5 instruction that is a combination of the previous instructions, is the **compare instruction**, or **CMP**. The CMP instruction is an input instruction that compares values from addresses or files. Figure 14–30 shows a CMP instruction and the compare expression is designated as T4:0.ACC = N7:2.

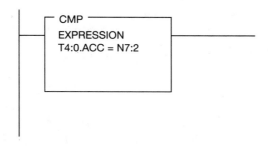

Figure 14–30 Allen-Bradley PLC-5 Compare Instruction

The CMP instruction in this case is true only when the accumulated value of T4:0 is equal to the value found in integer file 7, word 2.

The table in Figure 14–31 shows the different operators (symbols) that the CMP instruction uses. Because standard computer keyboards do not have keys for not equal, less than or equal to, or a greater than or equal to, the CMP instruction uses variations for these three operators as shown in Figure 14–31.

OPERATOR	DESCRIPTION	EXAMPLE
=	Equal to	True if A = B
< >	Not Equal to	True if A < > B
<	Less than	True if A< B
< =	Less than or equal to	True if A < = B
>	Greater than	True if A > B
> =	Greater than or equal to	True if A > = B

Figure 14–31 Available Operators (Symbols) for CMP Instruction

Figure 14–32 compares the CMP instruction to its companion data compare instructions.

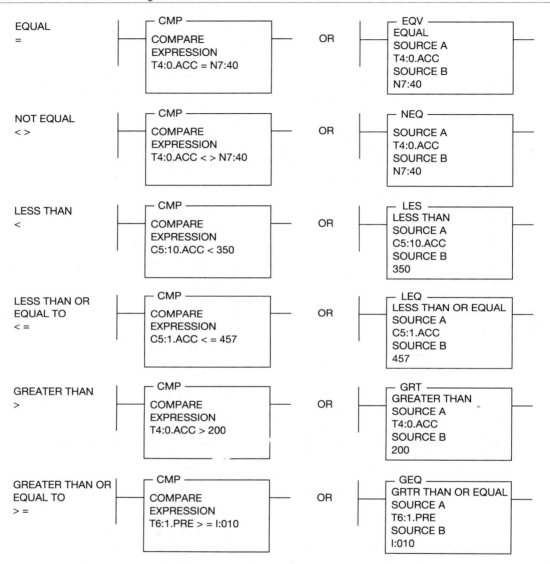

Figure 14–32 Comparing the PLC-5 CMP Instruction to Another PLC-5

NOTE: While the CMP instruction duplicates the other data compare instructions, the execution time for the CMP instruction is longer than the execution time for equivalent comparison instructions (for example, GRT, LEQ, etc.). A CMP instruction also uses more words per instruction than the equivalent comparison instructions. The advantage, however, is that the CMP instruction does all of the compare functions that are needed, without remembering all the instruction mnemonics.

The LIM, or limit test instruction, is an input instruction used by the PLC-5 to test for values inside or outside a specific range. The instruction is false until it detects that the test value is within certain limits. It then goes true. When the instruction detects that the test value has again gone outside the prescribed limits, the instruction goes false. This instruction is perfect for monitoring analog signals and making program decisions based on the analog value(s). Figure 14–33 shows the format for a LIM instruction.

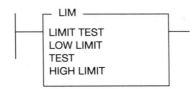

Figure 14–33 PLC-5 Limit Test Instruction Format

To program a LIM instruction, the following information must be provided:

Low Limit
Low limit is a constant, or an address that determines the lower limit of the test range. The value in the low limit can be either integer (whole number) or floating point (number[s] and a decimal).

Test Value
The test value is the address that contains the value examined to determine if it is inside or outside the specified range.

High Limit
High limit is a constant (numeric value) or an address that determines the upper limit of the test range. The value in the high limit can be either integer or floating point value.

An example of how a LIM instruction is used is shown in Figure 14–34. In the example, the LIM instruction is used to turn *ON* an indicator lamp whenever the accumulated value of T4:1 is between the values stored in N7:10 and N7:20.

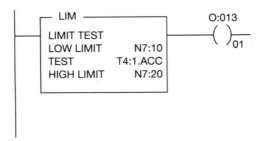

Figure 14–34 PLC-5 Limit Test Instruction
Monitoring Analog Output O:010

If the lower limit was set to 100 (value stored in N7:10) and the upper limit was set for 300 (value stored in N7:20), the instruction will be false as long as the accumulated value of T4:1 was less than 100 or greater than 300. When the accumulated value of T4:1 reached 100 and became equal to or greater than the low limit, the instruction would become true and indicator lamp (output O:013/01) would be set to 1, or turned *ON*. If the accumulated value of T4:1 became greater than 300, the instruction would again go false and the indicator lamp (O:013/01) would be turned *OFF*. The instruction remains *OFF* as long as the accumulated value of T4:1 is outside the limit test range (100-300) that was established for the LIM instruction.

The values that are used for the low limit and high limit can be entered as numeric values when the instruction is being programmed. Instead of referencing N7:20 (which held a value of 300), a value of 300 could have been entered at the low limit prompt.

Once the electrician or technician becomes familiar with a specific PLC, the many applications and advantages of using the various data compare instructions become evident.

Chapter Summary

Although formats and instructions vary with each PLC manufacturer, the concepts of data manipulation are the same. Data manipulation enables an operator to transfer data from one word location to another, whereas data comparison allows a value in one word to be compared to another word or a constant value.

Both data transfer and data comparison instructions give new dimension and flexibility to motor-control circuits, and the application of either is only limited by operator imagination.

REVIEW QUESTIONS

1. Define the term *data transfer*.
2. When numerical information replaces data that already exists in a memory location it is referred to as:
   a. exchanging info (data)
   b. replacement programming
   c. blanket move
   d. writing over
3. Match the symbols to their correct definitions.
   a. > _____ 1. less than 1
   b. < _____ 2. less than
   c. = _____ 3. less than or equal to
   d. ≥ _____ 4. greater than
   e. ≠ _____ 5. greater than or equal to
   f. <> _____ 6. equal to
   g. >= _____ 7. not equal to
   h. <= _____ 8. not equal to 1
   i. ≤ _____ 9. greater than 1

4. Define the term *data compare.*
5. Write a program that compares the accumulated value of T4:0 and constant of 250. The instruction is to be true when the accumulated value of T4:0 is greater than 250.
6. Define the term *mask.*
7. Give an example of how an Allen-Bradley LIM instruction is used.

Chapter 15

Math Functions

Objectives

After completing this chapter, you should have the knowledge to
- List the four standard math functions available with most PLCs.
- Discuss the math functions and give examples of how they are used.
- Express negative numbers in 2s complement.
- Add signed numbers.
- Convert a negative binary display to its decimal equivalent.
- Complete a subtraction problem using 2s complement and addition.

Using Math Functions

The four basic math, or arithmetic, functions are addition (+), subtraction (–), multiplication (×), and division (÷), and can be used with constant values or values stored in a storage register, holding register, input/output registers, or any other accessible word locations.

A typical application of an arithmetic, or math function may be a chemical batch plant where a given mix of two chemicals (A and B) is to have a 2:1 ratio. By using analog input devices (discussed in Chapter 2), the weight of the first chemical, A, can be converted to a binary equivalent and stored. For a 2:1 mix, only one half as much of chemical B (by weight) is needed. The binary value of the weight that is stored can be divided by two to determine the amount of chemical B that is needed for a proper 2:1 mix.

Figure 15–1, using a Square D *LET* instruction, shows how the value of chemical A (storage word S40) can be divided by 2, and the result placed in storage word S41. A data comparison is then made to the data in word S41 to limit the amount of chemical B to one half the amount of chemical A.

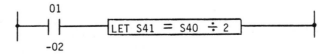

Figure 15–1 Square D Arithmetic Instruction

When 01-02 is closed, the value in word S40 (chemical A) is divided by 2, and the result is stored in word S41 on each processor scan. Once the total weight of chemical A has been determined and the amount divided by 2, the result (which is now stored in S41) is programmed with a data compare instruction to control the feed of chemical B. Figure 15–2 shows how an *IF* data compare instruction is programmed to control the amount of chemical B that is to be mixed with chemical A.

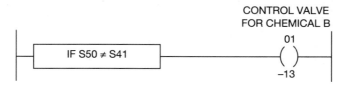

Figure 15–2 Programming an *IF* Data Compare
Instruction to Control the Amount of Chemical B
Used in the 2:1 Mix

As shown in the figure, output O1-13, which is the control valve for chemical B, is true, or *ON*, until the value in S50 equals S41. Storage word 50 (S50) is the word that stores the weight of chemical B. When the weight of chemical B is equal to one half the weight of chemical A (the value stored in S41), the *IF* instruction goes false, and the control valve to chemical B is closed.

The math functions might also be used with timer and/or counter values, accumulated or preset, to change a given process machine operation under varying conditions.

Using a Square D Company *LET* instruction, the basic arithmetic functions used with storage words and constants are shown in Figures 15–3a, 15–3b, 15–3c, and 15–3d.

Figure 15–3a Square D Add Instruction

When 01-08 is true, a constant of 003 is added to the value in storage word 20, and the sum is stored in storage word 15.

Figure 15–3b Square D Subtract Instruction

When 01-07 is true, the value in storage word 4 is subtracted from a constant of 1000, and the difference is stored in storage word 40.

Figure 15–3c Square D Multiply Instruction

When 01-03 is true, the value in storage word 18 is multiplied by the value of storage word 19, and the product is stored in storage word 16.

Figure 15–3d Square D Divide Instruction

When 01-05 is true, the value in storage word 18 is divided by 3, and the result is stored in storage word 30.

Math functions can also be used with Square D *IF* data comparison instructions as shown in Figure 15–4.

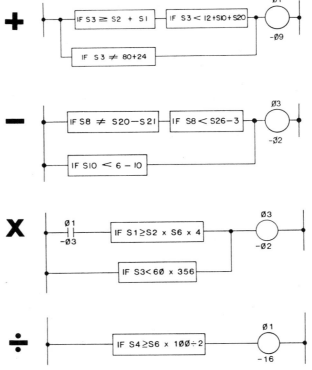

Figure 15–4 Combining Math and Data Comparison Instructions

Square D SY/MAX Model 300 uses a 16-bit word format for arithmetic functions, with bit 16 designated as the signed bit. The words used for the arithmetic functions may be values stored in registers, constants, or combinations of register values and constants.

Arithmetic Functions Using the Allen-Bradley PLC-2

The addition and subtraction instructions for the Allen-Bradley PLC-2 are used in conjunction with the *GET* instruction, and are examples of **single precision arithmetic**. Single precision arithmetic, or math, means that only one word is used to store the results of either addition or subtraction. When only one word is used to store the results, the number is limited in size, and in the PLC-2, the maximum size of a number is 999 for addition and subtraction. When using a technique called **double precision arithmetic**, two words are used to store the results of arithmetic operations, thereby increasing the size of the value that can be stored. The PLC-2 uses this technique for division and multiplication.

Figure 15–5 shows two *GET* instructions followed by the **ADDITION instruction**. When input device 11105 is true, the values stored in words 030 and 031 are added and the sum is stored in word 032. Allen-Bradley PLC-2 arithmetic functions limit the number in any storage word to 999 BCD. If the results of the addition exceed 999, bit 14 of the addition storage word (032) is set to 1, and a 1 is displayed on the programmer VDT.

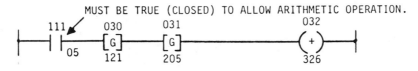

Figure 15–5 Allen-Bradley PLC-2 Add Instruction

The **SUBTRACTION instruction** uses two *GET* instructions (050 and 062) with a subtraction instruction (071) for single precision math as shown in Figure 15–6. If the result of the subtraction is a negative number, bit 16 of the subtraction storage word (071) is set to 1, and a minus sign (–) is displayed on the programmer video display terminal.

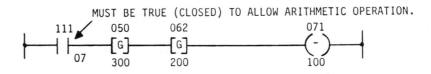

Figure 15–6 Allen-Bradley PLC-2 Subtract Instruction

The **DIVIDE** and **MULTIPLY instructions** for the Allen-Bradley PLC-2 family use two data table, or two storage words, to store the double precision results of both division and multiplication. The product of multiplication is limited to 998,001 (999 × 999) while the quotient of the division instruction is limited to three whole numbers and three decimal places. Figures 15–7 and 15–8 are examples of programmed multiply and divide instructions using the Allen-Bradley PLC-2 format.

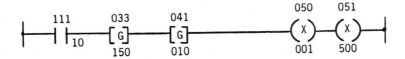

Figure 15–7 Allen-Bradley PLC-2 Multiply Instruction

When input device 11110 is true (closed), the value of word 033 (150) is multiplied by the value in word 041 (010), and the product (1500) is stored in words 050 and 051 (001 and 500, or 1500).

NOTE: Two memory words are used so that a number larger than 999 can be displayed and stored. This is referred to as double precision arithmetic.

Figure 15–8 Allen-Bradley PLC-2 Divide Instruction

When input device 11106 is true (closed), the value in word 047 (050) is divided by the value in word 054 (025), and the result is stored in words 032 and 033 (002.000, or 2).

Allen-Bradley PLC-5 Math Functions

For their PLC-5 family, Allen-Bradley uses a **compute (CPT) instruction**. The CPT instruction is an output instruction that performs the operations defined in the expression and then writes the results to the destination address. Figure 15–9 shows the CPT instruction format.

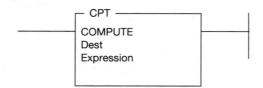

Figure 15–9 PLC-5 Compute (CPT) Instruction Format

The typical math functions that can be performed with the CPT instruction are add (+), subtract (–), multiply (*), divide (|), negate (–), square root (SQR), and exponential (**). The table in Figure 15-10 shows the functions and an example of each operation. The complexity of the math performed will vary with the PLC-5 model.

NOTE: Different symbols are used for some operations due to limited symbols on the keyboard.

Operator	Description	Example
+	Add	2 + 3 + 7
–	Subtract	12 – 5
*	Multiply	6 * (5 * 2)
|	Divide	24 | 4
–	Negate (negative)	–N7:0
SQR	Square Root	SQR N7:1
**	Exponential	10 ** 3 or 10^3

Figure 15–10 PLC-5 Compute (CPT) Operators (Symbols)

Figure 15–11 shows how a CPT instruction is typically programmed. In some PLC-5 models, but not all, any combination of operators may be used with various addresses and/or constants. The expression can be up to 80 characters in length.

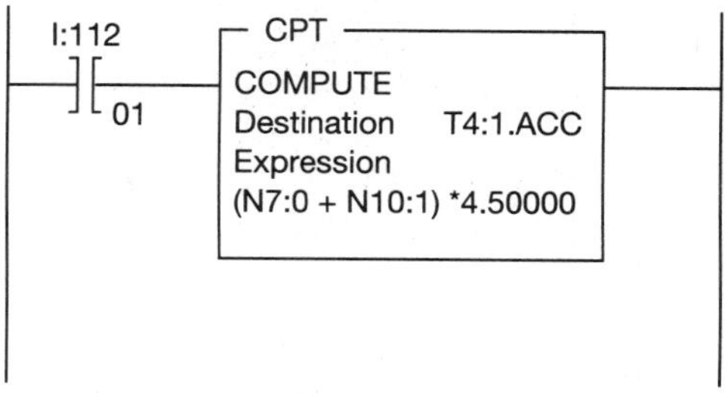

Figure 15–11 Programmed PLC-5 Compute (CPT) Instruction

When the instruction first appears on the screen during programming, the destination must be specified first. The destination address in the example is T4:1.ACC. Then the expression or math formula is entered. The expression states that the value in N7:0 is to be added to the value found in N10:1. The sum is then multiplied by 4.5. The result of the computation is then sent to the designated destination, in this case, the timer file, file 4, timer 1, accumulated value (T4:1.ACC).

Gould 984 Arithmetic (Math) Instruction

When using Modsoft® software for programming, pressing the Calculations (CALCS) designated software label key will provide the *ADD*, *SUB*, *MUL*, and *DIV* block functions. Figure 15–12 shows the format for the 984 *ADD* block function.

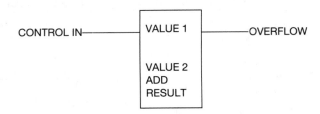

Figure 15–12 Format for Gould 984 Add Block Function

The top node is the first value (value 1), and can be a decimal number ranging from 1–999 in 16-bit CPUs or 1–9999 in 24-bit CPUs, an input register (30XXX), or a holding register (4XXXX).

The middle node is the second value (value 2), and can be a decimal number ranging from 1–999 in 16-bit CPUs or 1–9999 in 24-bit CPUs, an input register (30XXX), or a holding register (4XXXX).

The bottom node contains the symbol for the math function that is to be performed (*ADD*) and the address of the holding register where the results of the arithmetic will be stored (4XXXX).

The input line (control in) controls the block. If the control line is true, the math function is performed on each scan of the processor, and the results are stored in the designated holding register. The overflow line, top right, passes power, or is true, if the results of the math operation is a number higher than 9999. Figure 15–13 shows how the *ADD* function is programmed.

In Figure 15–13, when 10001 is closed, value 1 (50) is added to value 2 (100), and the result (150) is stored in register 40160.

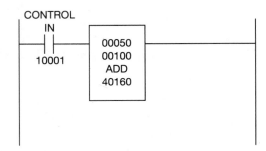

Figure 15–13 Add Function Block for Gould 984

The *SUB* function performs subtraction of values 1 and 2, and stores the result in a holding register. The function block holds the same information as the *ADD* block. The difference is the output side of the instruction. The block compares the values and identifies whether:

value 1 is > (greater than) value 2
value 1 is = (equal to) value 2
or if value 2 is > (greater than) value 1; this is the same as saying value 1 is < (less than) value 2.

Figure 15–14a shows a *SUB* function block, and Figure 15–14b shows how the block is programmed.

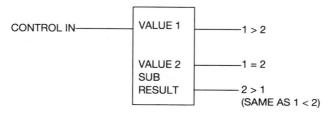

Figure 15–14a Subtract Function Block for Gould 984

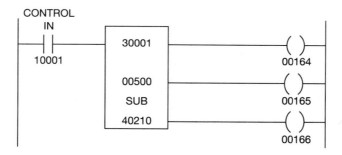

Figure 15–14b Programmed Subtract Function Block

When input device 10001 in the control in line is true, value 2 (500) is subtracted from value 1 (input register 30001), and the result is stored in holding register 40210 on each scan of the processor. Any time input 10001 is set to 1, or *ON*, the function block makes the following comparisons:

If the value in register 30001 is greater than (>) than 500, output 00164 is *ON*.

If the value in register 3001 is equal to (=) 500, output 00165 is set to 1, or turned *ON*.

Any time the value in register 30001 is less than (<) 500, output 00166 is turned *ON*.

The *MUL* (multiply) function multiplies value 1 by value 2 and stores the result in a holding register. The *MUL* function block is shown in Figure 15–15.

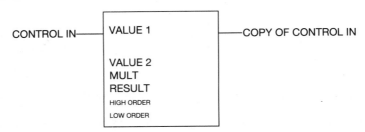

Figure 15–15 Gould 984 Multiply Function Block

In the *MUL* instruction, the bottom node contains the *MUL* symbol and a holding register (4XXXX). The results of the multiplication are stored in two consecutive registers (the same technique used for the PLC-2 family). The higher order digits are stored in the register specified in the bottom node, and the lower order digits are stored in the next sequential register. If 40001 is specified in the bottom node, register 40002 is used to store the lower order digits: Figure 15–16 shows how a *MUL* function block is programmed.

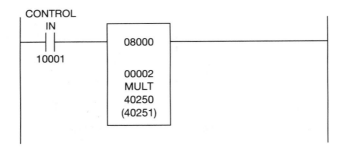

Figure 15–16 Gould 984 Multiply Function Block

When input device 10001 is closed, the block is activated and value 1 (8000) is multiplied by value 2 (2). The results are stored in two sequential registers. In this example, 40250 contains the higher order digits (0001) and 40251 contains the lower order digits (6000). The output line (top right) is true any time the control in line is true.

The last of the arithmetic functions is the *DIV*, or division block. The *DIV* function divides value 1 by value 2 and stores the result and remainder in two consecutive holding registers (4XXXX). Figure 15–17 shows the *DIV* block.

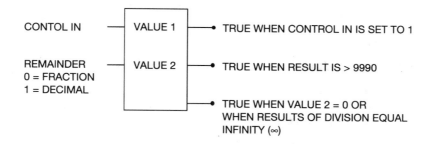

Figure 15–17 Division Block for Gould 984

The top node is value 1 and may be one of the following:
  a. a decimal number ranging from 1–999 in a 16-bit CPU and 1–9999 in a 24-bit CPU
  b. two sequential input registers. The first register holds the higher order numbers, and the second register holds the low order digits
  c. two sequential holding registers. The first register holds the higher order numbers, while the second register holds the low order digits.

The middle node is value 2 and can be:
  a. a decimal number ranging from 1–999 in a 16-bit CPU and 1–9999 in a 24-bit CPU
  b. an input register (30XXX) or a holding register (4XXXX).

The bottom node contains the symbol *DIV* and the designated holding register. The results of the division are stored in the designated holding register, while the remainder is stored in the next sequential register.

The inputs function as follows:

Top: Control In line. When the control in line is true, the processor performs the required division on each scan;

Middle: When this line is true, the remainder is expressed in decimal form. When this line is false, the remainder is expressed as a fraction.

The outputs are as follows:

Top: True when control in line is true;

Middle: True when the result of the division is > than 9999;

Bottom: True when value 2 = 0 (zero) or if the result of the division would be infinity (∞).

Figure 15–18 shows how the *DIV* block is programmed.

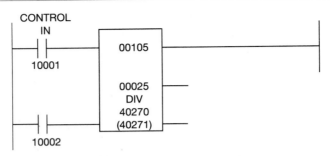

Figure 15–18 Programmed DIV Function

When 10001 is closed, value 1 (105) is divided by value 2 (25) and the results (4, with a remainder of 2) is stored in two sequential registers. Register 40270 contains the whole number and register 40271 contains the remainder. If 10002 is open, the remainder is fractional (5/25 in this example). If 10002 is closed, the remainder is expressed as a decimal (0.2 in this example). Register 40271 displays 20000. Although no decimal is shown, the decimal is implied because register 40271 holds the remainder.

The limits and limitations of the different arithmetic functions vary from manufacturer to manufacturer, but the concepts are basically the same. Values or contents of storage registers, data table words, or constants are combined arithmetically, and the results are stored in registers or data table words. Although the actual arithmetic function is performed using binary 2s complement, the values may be stored and/or displayed in binary, BCD, hexadecimal, or octal numbering systems, depending on the PLC. The only way to learn how to apply the arithmetic functions on a given PLC is to read the manufacturer's literature and *work* with *that* PLC.

2s Compliment

Virtually all programmable controllers, computers, and other electronic calculating equipment perform counting functions using the binary system. For those PLCs that are programmed to perform arithmetic functions, a method of representing both positive (+) and negative (−) numbers must be used. The most common method is 2s complement. The 2s complement is simply a convention for binary representation of negative decimal numbers.

Before going any further with a discussion of 2s complement, a review of adding binary numbers may be helpful. In decimal addition, numbers are added according to an addition table. A partial addition table is shown in Fig. 15–19.

	0	1	2	3	4	5
0	0	1	2	3	4	
1	1	2	3	4	5	
2	2	3	4	5	6	
3	3	4	5			
4	4	5				

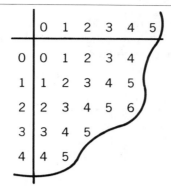

Figure 15–19 Decimal Addition System

To use the table, the first number to be added is located on the vertical line, and the second number on the horizontal line. The sum, or total, is found where the two imaginary lines intersect. For example, 3 + 2 = 5 (as shown in Figure 15–20).

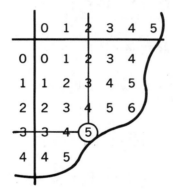

Figure 15–20 Adding 2 and 3

For binary addition, a similar addition table is constructed. The table is small because the binary system only has two digits (1 and 0) (Figure 15–21).

	0	1
0	0	1
1	1	10

Figure 15–21 Binary Addition Table

To use the table, the first number (digit) to be added is located on the vertical line, the second digit is located on the horizontal line. The sum, or total, is found where the two immaginary lines intersect. Figure 15–22 shows an example of adding $1 + 0 = 1$.

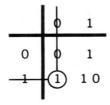

Figure 15–22 Adding Binary 1 and 0

Notice that if 1 and 1 are added, the table shows 1 0, not 2, as might be expected. 1 0 is the binary representation of 2 (Figure 15–23).

$$\begin{array}{|c|c|c|c|} \hline 8 & 4 & 2 & 1 \\ \hline 0 & 0 & 1 & 0 \\ \hline \end{array}_2$$

Figure 15–23 Binary Representation of 2

Figure 15–24 shows how binary numbers 1011_2 and 110_2 are added.

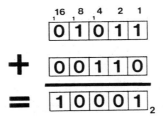

Figure 15–24 Adding Binary Numbers

In the 1s column $1 + 0 = 1$.
In the 2s column $1 + 1 = 0$ with a carry over of 1.
In the 4s column $1 + 0 + 1 = 0$, with a carry over of 1.
In the 8s column $1 + 1 + 0 = 0$, with a carry over of 1.
In the 16s column $1 + 0 + 0 = 1$.
The sum (total) of 1011_2 and 110_2 is, therefore, 10001_2.

To verify our results we can convert the binary numbers to decimal equivalent numbers and add them as shown in Figure 15–25.

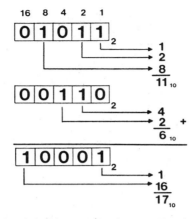

Figure 15–25 Converting Binary Numbers to Decimal Equivalents

Another example of adding binary numbers is shown in Figure 15–26 when 11011_2 and 11_2 are added.

$$\begin{array}{r} {\scriptstyle 1\ 1}\\ 11011 \\ +\ \ \ \ 11 \\ \hline 11110_2 \end{array}$$

Figure 15–26 Addition of Binary Numbers

In the 1s column $1 + 1 = 0$, with a carry over of 1.
In the 2s column $1 + 1 + 1 = 1$, with a carry over of 1.

> **NOTE:** $1 + 1 + 1 = 3$. The binary equivalent of 3_{10} is 11_2.

In the 4s column $1 + 0 + 0 = 1$. In the 8s column $1 + 0 = 1$.
In the 16s column $1 + 0 = 1$.
The sum of 11011_2 and 11_2 is 11110_2.

To verify this method, convert the binary numbers to decimal numbers, and add them as shown in Figure 15–27.

$$\begin{array}{r} 11011_2 = 27_{10} \\ +\ \ \ \ 11_2 = 3_{10} \\ \hline 11110_2 = 30_{10} \end{array}$$

Figure 15–27 Comparing Binary and Decimal Addition

To represent negative numbers using the binary numbering system, one bit is designated as a **signed bit**. If the designated bit is a 0 (zero), the number is positive, and if the bit is a 1, the number is negative.

Using a 4-bit word length, and using bit 4 as the designated signed bit, 0001_2 represents +1 decimal (see Figure 15–28).

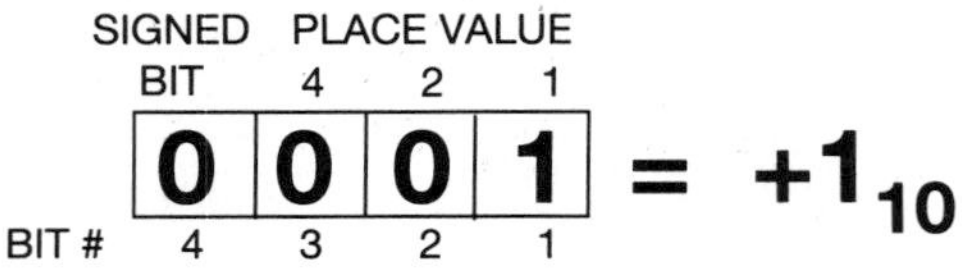

Figure 15–28 4-Bit Word with a Signed Bit

The table in Figure 15–29 shows all of the possible numbers for a 4-bit word using 2s complement.

Binary number	Decimal
0111	+7
0110	+6
0101	+5
0100	+4
0011	+3
0010	+2
0001	+1
0000	0
1111	–1
1110	–2
1101	–3
1100	–4
1011	–5
1010	–6
1001	–7
1000	–8

Figure 15–29 2s Complement numbers for an 8-Bit Word

Notice that the negative numbers go to –8 while the positive numbers only go to +7. In this case, the signed bit is used for its place value, which is 8. The same holds true for 8- and 16-bit words. The maximum negative number is always one number *higher* than the maximum positive number.

In Chapter 13, the preset and accumulated values for the Allen-Bradley PLC-5 counters was given as –32,768 to +32,767. The text indicated that the negative value was stored in 2s complement form. When using a 16-bit word with a signed bit, the highest positive number that can be obtained is 32,767, whereas the highest negative number is –32,768. Figure 15–30 shows 16-bit words displaying the maximum positive and negative numbers.

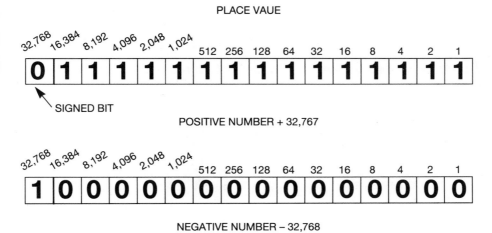

Figure 15–30 Highest Possible Positive and Negative Numbers Using a 16-Bit Word and 2s Complement

To display a negative binary number requires that the same value positive number be complemented (all 1s changed to 0s and all 0s changed to 1s) and a value of 1 added. The result is 2s complement of the number. Figure 15–31 shows the steps to express –5 in 2s complement using a 4-bit word.

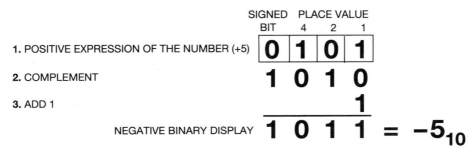

Figure 15–31 Expressing –5 in 2s Complement

Another example of 2s complement is shown in Figure 15–32 with the steps required to express –7 in 2s complement.

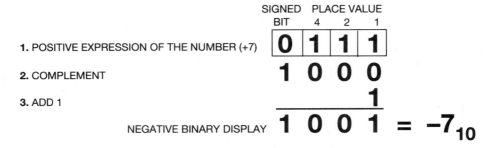

Figure 15–32 Expressing –7 in 2s Complement

To convert a negative binary number to the decimal equivalent, the negative binary display is complemented, 1 is added, the binary sum is converted to decimal, and the negative sign (–) is added. Figure 15–33 shows what steps are necessary to determine the negative value of 1110.

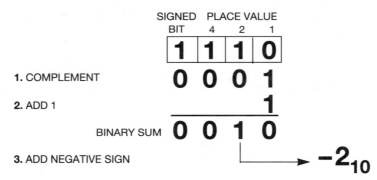

Figure 15–33 2s Complement to Decimal Equivalent

Another method of converting positive numbers to 2s complemented negative numbers is as follows: starting at the least significant bit and working to the left, copy each bit up to and including the first 1 bit, and then complement or change each remaining bit. Figure 15–34 shows this alternate method of expressing –2 in 2s compliment using a 4-bit word.

Figure 15–34 Alternate Method of 2s Complement

A further example is shown in Figure 15–35 for 2s complementing the value 24 using an 8-bit word.

1. ORIGINAL POSITIVE NUMBER (+24)	0	0	0	1	1	0	0	0
2. COPY UP TO FIRST 1 BIT					1	0	0	0
3. COMPLEMENT REMAINING BITS	1	1	1	0	1	0	0	0
2S COMPLEMENT –24	1	1	1	0	1	0	0	0

Figure 15–35 2s Complement of –24 Decimal

By using 2s complement, negative and positive values can now be added. The two steps for adding -7_{10} and $+5_{10}$ using 2s complement with a 4-bit word is shown in Figure 15–36.

		SIGNED BIT	PLACE VALUE 4	2	1
1. EXPRESS –7 IN 2s COMPLEMENT					
a. POSITIVE EXPRESSION OF NUMBER (+7)		0	1	1	1
b. COMPLEMENT		1	0	0	0
c. ADD 1					1
NEGATIVE BINARY DISPLAY		1	0	0	1
2. ADD –7 AND +5		1	0	0	1
	+	0	1	0	1
BINARY SUM		1	1	1	0 = -2_{10}

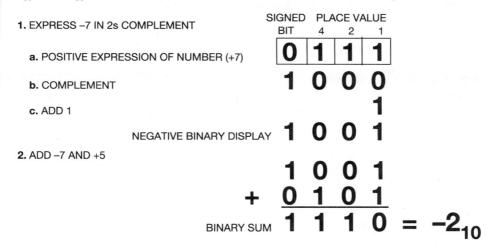

Figure 15–36 Adding Positive and Negative Numbers

NOTE: When adding signed binary numbers, any carry over from the signed bit column is discarded.

If a Square D *LET* statement is programmed to perform the addition in the previous example, the rung appears as shown in Figure 15–37.

```
!
!I0001                                                       +--------------------------+
+--] [-+-------+-------+-------+-------+-------+--------!LET    S0018 =S0019 +S0020!
!  -07                                                       +------  0000  -0007    0005
```

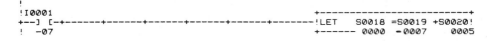

Figure 15–37 Programmed Addition of Negative and Positive Numbers

NOTE: To enter a negative value into a storage register with the SY/MAX 300, place a zero after the equal sign and subtract the required number from it. For example, *LET* S19 = 0 – 7. This programming technique enters –7 into register S19.

When input device 01-07 is closed, or goes true, the values stored in registers S19 and S20 (–7 and +5) will be added, and the sum (–2) stored in register S18.

Using the data mode of the SY/MAX CRT, the register display for registers 18, 19, and 20 (S18, S19, and S20) appear as indicated in Figure 15–38.

```
              DECIMAL    HEX    16... .BIN/FLP. ...1      32... .STATUS.. ...17
    S0019      -0007     FFF9   1111 1111 1.111 1001      0001 0011 0000 0000
    S0020       0005     0005   0000 0000 0000 0101       0001 0011 0000 0000
    S0018      -0002     FFFE   1111 1111 1111 1110       0001 0011 0000 0000
```

Figure 15–38 VDT Display of Registers Showing Decimal, Hexadecimal and Binary Values

Once addition of signed numbers is possible, the other arithmetic functions (subtraction, multiplication and division) are also possible, because they are achieved by successive addition on a PLC.

EXAMPLE: Subtracting the number 20 from 26 is accomplished by complementing 20 to obtain –20, and then performing addition.

Subtracting 20 from 26 by complementing 20 and performing addition using an 8-bit word is shown in Figure 15–39.

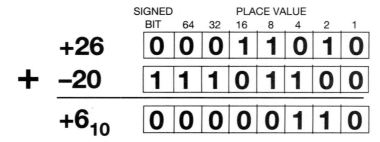

Figure 15–39 Subtraction by Addition

Chapter Summary

For programmable controllers to perform arithmetic functions, a way must be found to represent both positive and negative numbers. One of the most common methods used is called 2s complement. Using 2s complement, negative and positive numbers can be added, subtracted, divided, and multiplied. In reality, however, all arithmetic functions are accomplished by successive addition. As with most other features, arithmetic functions and formats vary with each manufacturer. Arithmetic functions of add, subtract, multiply, and divide can be combined with data manipulation instructions (data transfer and data compare) to provide expanded control and information from process or driven equipment. Memory words such as holding, storage, and data can be used with the arithmetic functions as well as words and constants, or just constants.

REVIEW QUESTIONS_____

1. List the four math functions that can be performed by most PLCs.
2. Express the following signed decimal numbers in 2s complement. Use 8-bit words. Show all work.
   a.   (–)7
   b.   (–)4
   c.   (–)3

3. Convert the following decimal numbers to 2s complement and add. Use 8-bit words. Show all work.
   a.   (+)4
        (–)7

   b.   (–)10
        (+)22

   c.   (+)22
        (+)22

4. T   F   Data manipulation instructions can be combined with arithmetic (math) instructions.
5. What is *double precision math* used for?
6. Give an example of how a math function is used in a PLC program.

Chapter 16

Word and File Moves

Objectives

After completing this chapter, you should have the knowledge to
- Describe the function of a synchronous shift register.
- Explain the function of *word-to-file*, *file-to-word*, and *file-to-file* instructions.
- Explain the difference between an *asynchronous shift register* (*FIFO*) and a *word-to-file* move.

Before word and file moves are discussed, the electrician and technician should understand the definition of both words and files.

Words, or registers as they are often referred to, are locations in memory that can be used to store different kinds of information. Typically, a word or register can store the status of inputs and outputs, hold numerical values used for math functions, other numerical data used for timers and counters, etc. Most words consist of 16 bits, although on older PLCs, an 8-bit word is sometimes used.

A **file** is a group of consecutive memory words used to store information. Words 1 through 5 would make up a consecutive 5-word file. Words 1, 2, 3, 6, and 7 are not used as a 5-word file because the numbers are not consecutive. A file is also referred to as a **table** by some PLC manufacturers.

Words

Information stored in a word can be shifted within the word, or from one word to another. Information stored in a word may also be moved into a file, or the information stored in a file can be transferred into a word. All of these different possibilities are discussed later in this chapter.

Synchronous Shift Register

When information is shifted—one bit at a time—within a word, or from one word to another, it is called a **synchronous shift register**. The bits may be shifted forward (left) or reverse (right).

NOTE: The synchronous shift register may also be referred to as a **serial shift register** or **bit shift**.

Figure 16–1 shows a 16–bit word used as a forward synchronous shift register.

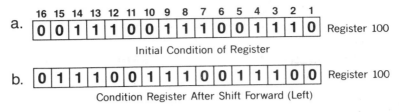

Figure 16–1 Forward 16-Bit Synchronous Shift Register

Figure 16-1 (a) shows the bit status of register word 100 prior to the forward shift, while 16–1 (b) shows how the register looks after the bits have been shifted one place to the left, or forward.

Notice that when the register is shifted, the information (1 or 0) in bit 16 is shifted out, and is lost. If the register is continually shifted with a zero (0) in bit location 1, all of the 1s are shifted left (forward) until only 0s remain (Figure 16–2).

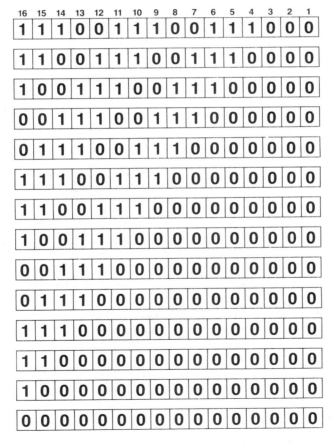

Figure 16–2 Register With All 1s Shifted Out

In a forward shift register, bit 1 is used to enter data (1 or 0). The data is then shifted forward, one bit at a time. Figure 16–3 shows a 2-word forward shift register. In this case data is entered at bit 1 and shifted one bit at a time to the left. With a 2-word shift register, the information (1 or 0) in bit 16 of word 1 is not shifted out and lost, but is shifted into bit 1 of the second word of the shift register.

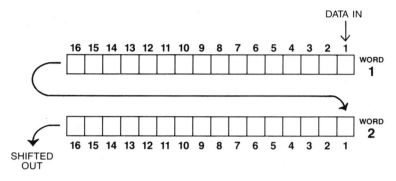

Figure 16–3 2-Word Forward Shift Register

Figure 16–4 shows the Square D Company format for a shift register.

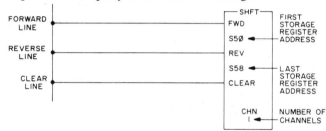

Figure 16–4 Square D Company Format for a Synchronous Shift Register
(Courtesy of Square D Company)

The shift register consists of a forward line, a reverse line, a clear line, and a shift box that contains three pieces of information. The information consists of the first storage register address to be used in the shift register, the last storage register address (which would be the same as the first storage register if the shift register used only one 16-bit word), and the number of channels. The channel specifies the number of bits to be shifted at one time. The choices are 1 as shown in the previous example, 8 or 16 bits at a time.

An 8-channel shift register shifts 8 bits at a time, either forward or reverse. A 16-channel shift register shifts all sixteen bits from one word to another. The forward line shifts the data in the register(s) forward (left) on each false-to-true (open-to-close) transition. The reverse line shifts the data in the register(s) reverse (right) on each false-to true-transition. The clear line enables the shift box when there is continuity. When the clear line is open, the data in the register(s) is set to 0, and transitions of either the forward or reverse lines have no effect.

An example of a practical application for a shift register is the overhead parts conveyor in Figure 16–5 which is used to transport parts into a paint booth for painting. If a part is on the hook as it enters the paint booth, limit switch 1 (LS-1) is activated. Limit switch 2 is activated each time a hook on the conveyor passes, even if no part is present.

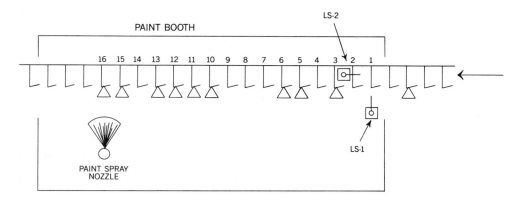

Figure 16–5 Applying a Forward Shift Register

Both limit switches (LS-1 and LS-2) are wired to an input module of a PLC, and the solenoid that operates the paint spray nozzle is wired to an output module, as shown in Figure 16–6.

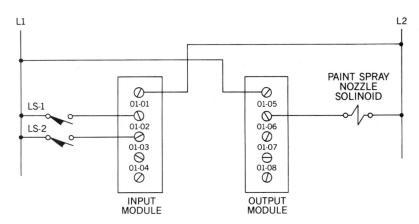

Figure 16–6 Input and Output Devices Wired to I/O Modules

The input addresses of the limit switches and the output address of the paint spray nozzle is now programmed with a forward shift register (Figure 16–7).

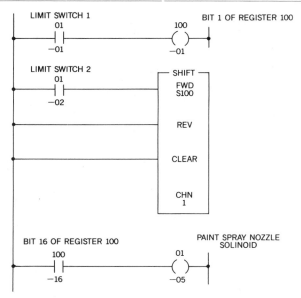

Figure 16–7 contents:

LIMIT SWITCH 1
01
┤ ├
−01

BIT 1 OF REGISTER 100
100
( )
−01

LIMIT SWITCH 2
01
┤ ├
−02

┌─ SHIFT ─┐
│ FWD   │
│ S100   │
│   │
│ REV   │
│   │
│ CLEAR │
│   │
│ CHN   │
│ 1   │
└────────┘

BIT 16 OF REGISTER 100
100
┤ ├
−16

PAINT SPRAY NOZZLE
SOLINOID
01
( )
−05

Figure 16–7 Programming a Foward Shift Register Using Square D Company Format

Notice that in the shift box in Figure 16–7, storage word 100 is being used as the shift register.

When a part activates LS-1 (address 01-01) and the limit switch closes, Rung 1 goes true, and a 1 is placed in bit 01 of word 100. As the part moves toward the spray nozzle, LS-2 is activated by the hook which closes LS-2. The closing of the LS-2 contacts (address 01-02) gives a false-to-true transition in the forward (FWD) line and the shift register shifts the information in word 100 one place to the left. The shift moves the 1 in the bit 1 location to the bit-2 location. As the conveyor continues to run, a 1 or 0 is entered into the bit 1 location of word 100, depending on whether a part is present or not. The data is then shifted as LS-2 is activated and deactivated by the moving hooks. As the data shifts to the left, the paint spray nozzle solenoid in Rung 3 is activated each time a 1 is shifted into bit 16 of word 100.

A reverse shift register is shown in Figure 16–8. Notice that with a reverse shift, the data enters at bit 16 and is shifted out at bit 1.

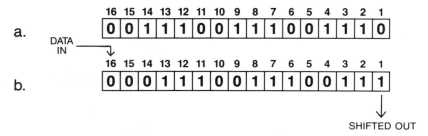

Figure 16–8 Reverse (Right) Shift Register

Similar to other functions, shift register formats vary from manufacturer to manufacturer, but the basic function of the synchronous shift register is the same.

File Moves

As indicated earlier, a file, or table, is a group of consecutive words used to store or hold information. A file can consist of just a few words or can be several hundred words in length, depending on the PLC. Figure 16–9 shows a 5-word file using consecutive memory words 50 through 54.

FILE

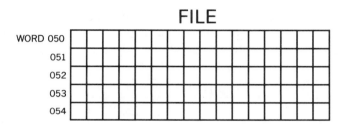

Figure 16–9 5-Word File

Information (data) may be transferred into or out of a file by using data transfer instructions. The three most common data transfer instructions are word-to-file, file-to-word, and file-to-file.

Word-to-File Instruction

The word-to-file instruction is used to transfer data from a word into a file. For example, word 110 stores the temperature of the die for a plastic injection molding machine. A thermocouple is attached to the heated die and then connected to a thermocouple input module. Depending on the module, the temperature of the die is then stored in an input word in either binary or BCD format. By using a word-to-file data transfer instruction, the data (temperature) in word 110 can be transferred into a file. Once the word-to-file instruction has been programmed, the information stored in word 110 is transferred into a file each time the instruction is implemented. Figure 16–10a shows a 5-word file prior to a word-to-file instruction being implemented, and Figure 16–10b shows the file after the data transfer instruction is implemented.

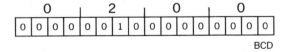

Figure 16–10a File Prior to Word-to-File Data Instruction Implementation

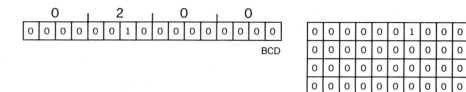

Figure 16–10b File Content After First Word-to-File Instruction is Implemented

The next time the word-to-file instruction is implemented (indexed), the current value in word 110 is transferred to the file. All information currently in the file is moved down to make room for the new data. Figure 16–11 illustrates this point.

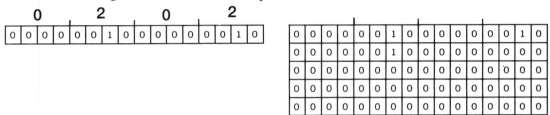

Figure 16–11 File After Second Word-to-File Instruction is Implemented

By using a timer, the word-to-file instruction could be implemented every 15 minutes. By increasing the size of the file, a record of the die temperature for an 8-hour shift can be stored, the data (temperature) from the file could be printed out, and the temperature of the die compared to quality-control records. The application of this instruction, like all other instructions, is limited only by imagination.

File-to-Word Instruction

The file-to-word instruction transfers data from a file into a word.

Using the previous example, the temperature of the injection molding machine die can be transferred to an output word (011) that controls a LED display. By incrementing or indexing the file-to-word instruction, the temperature of the die in 15 minute intervals is displayed. Figures 16–12a and 16–12b illustrate how a file-to-word instruction functions.

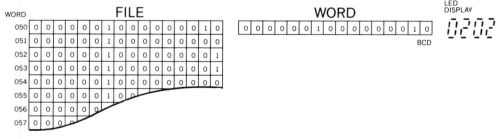

Figure 16–12a File-to-Word to Instruction at First Word of File (050)

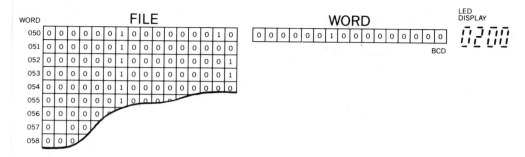

Figure 16–12b File-to-Word to Instruction at Second Word in File (051)

File-to-File Instruction

This instruction moves data from one file to another. The data from the source file may be moved to the destination file one word at a time, or the entire contents of the file can be moved in one move, depending on the PLC.

An example of using the file-to-file move might be a chemical batch plant where different amounts and types of chemicals are mixed for a variety of products. The different mix ratios (recipes) are stored in different files, and could be transferred to a file that controls machine and/or plant operation for a given product.

The file-to-file move is similar to a *GET* and *PUT* or *LET* instruction, except rather than just transferring data from one word to another, the file-to-file move transfers data from several words at one time. Figure 16–13 shows how a file-to-file move works.

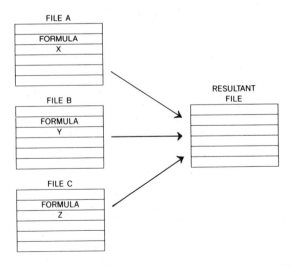

Figure 16–13 Data in File A, B, or C can be Transferred into the
Resultant File When the File-to-File Move is Executed

The Gould 984 PLC uses a *DX* (data transfer) function for making word-to-file, file-to-word, and file-to-file moves. Gould uses the term *register* rather than word, and *table* rather than file. Therefore, their *DX* instruction makes register-to-table, table-to-register, and table-to-table moves. Figure 16–14 shows the format for the Gould 984 data transfer instruction.

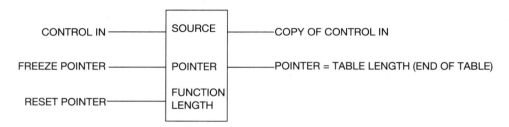

Figure 16–14 Gould 984 Data Transfer (DX) Instruction

The Control In line is used to activate the instruction. The middle line on the left is the **Freeze Pointer line**, and the bottom line on the left is the **Reset Pointer line**. The top line to the right is the **Copy of Control In** line; this line is true anytime that the control in line is true. This line can be used for cascading instructions. The middle line on the right side is the **Pointer** line, and keeps tract of the number of data moves. This line is true when the data moves are equal to the data table length.

Within the function block, the top node is the data source and can be one of the following:
  The first of 16 logic coils (OXXXX)
  The first of 16 inputs (1XXXX)
  An input register (30XXX)
  A holding register (4XXXX)

The middle node must be a holding register (4XXXX). This register is the pointer to the destination table which must be consecutive with this word. If the pointer register is 40300, the data table must start with register 40301.

The bottom node contains the symbol that indicates the type of instruction it is, and a number to indicate the length of the table. The length of the data table can range from 1–255 16-bit words on some models, and 1–999 on other 984 models.

Figure 16–15 illustrates how the *DX* instruction works and is programmed. The bottom node contains the symbol R→T, and indicates that the *DX* instruction is programmed to perform a register-to-table move.

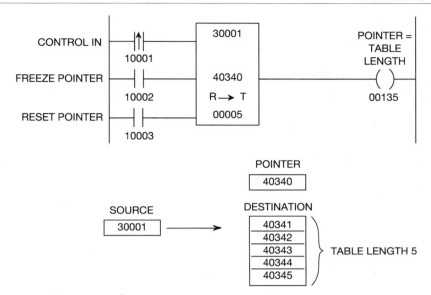

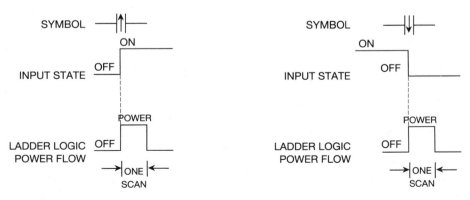

Figure 16–15 984 Data Transfer (DX) Instruction for Register to Table Moves

The contact in the Control In line is called a **positive transition contact**. When this instruction is programmed, the contact is only true for one scan when the input device makes an *OFF* to *ON* transition. This is similar to the Allen-Bradley PLC-5 one shot instruction. On the next *OFF* to *ON* transition, the instruction is again true, but only for one scan. Figure 16–16a shows the logic for the positive transitional contact. There is also a **negative transition contact** that is true for one scan on each *ON* to *OFF* transition of the input device. Figure 16–16b shows the negative transitional contact and the logic based on the state of the input.

Figure 16–16a Gould Positive Transitional Contact Logic

Figure 16–16b Gould Negative Transitional Contact Logic

With the first transition of input device 10001, the data in register 30001 is copied into the first word of the table (40341), and the pointer in pointer register 40340 is incremented to 1. On the second transition of 10001, the data in register 30001 is transferred into table word 40342 and the pointer in register 40340 is incremented to 2. With each successive transition of input device 10001, the data in register 30001 is sequentially moved into table words 40343, 40344, and 40345. Each transition also causes the pointer to increment from 2, to 3, to 4, and to 5. When the pointer value is equal to the table length (5 = 5), the center pointer line goes true, and output 00135 is turned *ON*.

It is possible to freeze the pointer at any location in the table. If the pointer is frozen at 3, the data from register 30001 is placed in table word 40343 (third word) with each transition of input device 10001. The pointer is frozen by closing input device 10002. The pointer is reset to 0 by activating the Reset Pointer line (input device 10003).

The 984 *DX* instruction also performs a table-to-register move (file-to-word) as shown in Figure 16–17.

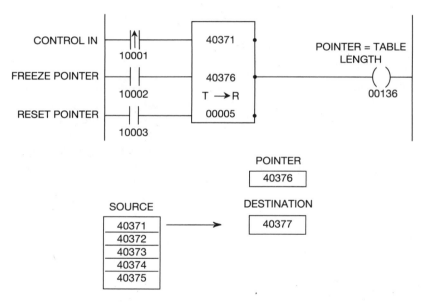

Figure 16–17 984 DX Instruction Programmed for a Table-to-Register Move

On each transition of input device 10001, data is moved sequentially from the table into the register (40377). On the first transition, the data in table word 40371 is copied, or moved, into register 40377, and the pointer is incremented to 1. On the next transition, the data from table word 40372 is transferred into register 40377 and the pointer incremented to 2. This process continues with each transition of 10001 until the last word in the table has been transferred into register word 40377. At this time, the pointer equals the table size, and output 00136 is turned *ON*. The pointer can be frozen at any point, controlling the data that is transferred into register 40377.

The *DX* instruction makes table-to-table moves in two ways. One table-to-table instruction moves one word at a time from the source table to the destination table. A second instruction transfers data from all words in the source table to the destination table in one move. Gould calls this latter instruction a **block move**. Figure 16–18 shows how the table-to-table move (T→T) transfers, or copies, the bit status of one word in the source table to the corresponding word in the destination file.

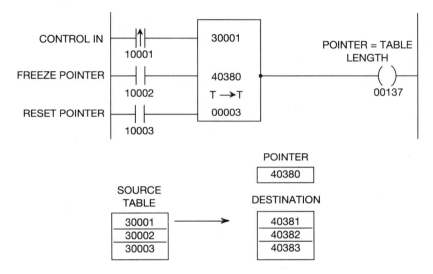

Figure 16–18 Table-to-Table Instruction for Gould 984

With the first transition of input device 10001, the instruction copies the bit status of 30001 into 40381. The pointer (40380) is incremented to 1. With each transition of 10001, data is transferred from subsequent source table words to the corresponding destination word. Although the table only consists of three words in the example, the table length could be up to 999 words, depending on the model of processor. As in previous *DX* instructions, if input device 10002 is turned *ON*, the pointer is frozen and the table-to-table move stops at whatever position it is when the pointer is frozen. To reinitiate the T→T function, the pointer is reset by closing and then opening input 10003. Once the pointer is reset, transitions of 1001 again cause data to be transferred from the source table to the destination table.

The Gould **block move (BLKM) instruction** moves all the data in one file to another file in one move. Figure 16–19 shows how the block move is programmed.

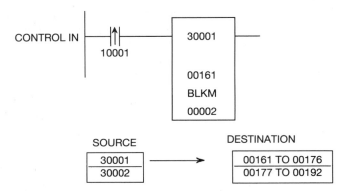

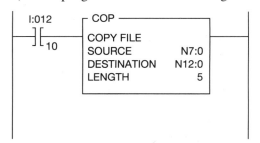

Figure 16–19 Gould 984 Block Move (BLKM) Instruction

The top node is the source table and displays the first word of the source table (30001). The center node is the destination table and displays the first word in the destination table. In this example, the destination table is a table of outputs (16 bits, 00161–00176). On each transition of input 10001, the contents of the source file is transferred into the destination file. This instruction does not have a pointer register because the data is not transferred a word at a time, but is done completely in one move. Although this example uses an output address for the destination table, the table could have consisted of holding registers. Several BLKM instructions can be used to transfer various formulas (data) as illustrated in Figure 16–13.

The Allen-Bradley PLC-5 uses a **file copy (COP) instruction** for making file-to-file moves. File copy is an output instruction, and is programmed as shown in Figure 16–20.

```
   I:012   ┌ COP ─────────────────────────┐
    ┤ ├    │ COPY FILE                     │
      10   │ SOURCE          N7:0          │
           │ DESTINATION     N12:0         │
           │ LENGTH             5          │
           └───────────────────────────────┘
```

Figure 16–20 PLC-5 File Copy (COP) Instruction

The address of the first word in the source file is specified as well as the first address of the destination file. The length of the file is then specified. In this example, the source file starts with N7:0, the destination file is N12:0, and the file length is 5 words. When the input device is closed, the *COP* instruction is enabled, and the instruction copies data from the 5-word file starting at N7:0 into the 5 words of the destination file starting at N12:0.

When using the *COP* instruction, it is important to note that if the destination is a file of words, such as an integer file, the programmer must specify the file length in words. If, however, the source file is a timer or counter address, the table length is specified for the number of timers and/or counters, not words. Remember that each timer and/or counter instruction requires three words. The destination file automatically adjusts in length when the source file is a timer or counter address. For example, if the source file contains 5 timer and/or counter addresses, the destination table is automatically set to 15 words.

Asynchronous Shift Register (FIFO)

The **asynchronous shift register**, instead of shifting bits of information within a word, or words, like the synchronous shift register, shifts the data from a complete word into a file, or stack. Although this appears to be just another name for a word-to-file instruction, it is not. There are similarities between the two, but there is also one major difference. In a word-to-file move, the information from the word is shifted into the top of the file and moved down through the file with each implementation, or indexing, of the instruction. The asynchronous shift register, however, allows the information transferred from a word to go to the last *unused* word of the file. This difference is why the asynchronous shift register is often referred to as a **FIFO stack** (first in–first out). Figure 16–21 compares the asynchronous shift register to the word-to-file instruction to demonstrate the difference.

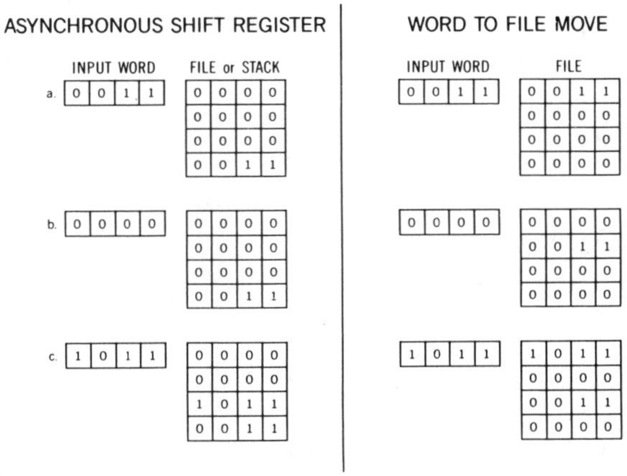

ASYNCHRONOUS SHIFT REGISTER | WORD TO FILE MOVE

INPUT WORD   FILE or STACK   INPUT WORD   FILE

a. 0 0 1 1 | 0 0 0 0 / 0 0 0 0 / 0 0 0 0 / 0 0 1 1   0 0 1 1 | 0 0 1 1 / 0 0 0 0 / 0 0 0 0 / 0 0 0 0

b. 0 0 0 0 | 0 0 0 0 / 0 0 0 0 / 0 0 0 0 / 0 0 1 1   0 0 0 0 | 0 0 0 0 / 0 0 1 1 / 0 0 0 0 / 0 0 0 0

c. 1 0 1 1 | 0 0 0 0 / 0 0 0 0 / 1 0 1 1 / 0 0 1 1   1 0 1 1 | 1 0 1 1 / 0 0 0 0 / 0 0 1 1 / 0 0 0 0

Figure 16–21 Comparison of Asynchronous
Shift Register (FIFO) to Word-to-File Move

When data is transferred at position (a), the asynchronous shift register places the data in the last (bottom) unused word of the file or stack. In the word-to-file move, however, data is transferred into the first (top) word of the file.

At the next implementation at position (b), no data is present in the input word, and no change takes place in the FIFO stack, but the previous data is moved *down* one word in the word-to-file instruction.

When the instructions are indexed again at position (c), the data of the input word is transferred to the next available word at the bottom of the FIFO stack, whereas the data is entered at the top of the file and all previous data is shifted *down* in the word-to-file instruction.

The biggest difference between the two instructions is the ability of the word-to-file move to accept and store all zeros, whereas the asynchronous shift register (FIFO) ignores input data of all zeros. The FIFO stack is well suited when the programmer is only interested in data, and not concerned with periods where no data (all zeros) is transferred.

The Allen-Bradley PLC-5 uses a pair of output instructions to store and retrieve data in a prescribed order, or FIFO. Figure 16–22 shows the **first in–first out load (FFL)** and **first in–first out unload (FFU) instructions**.

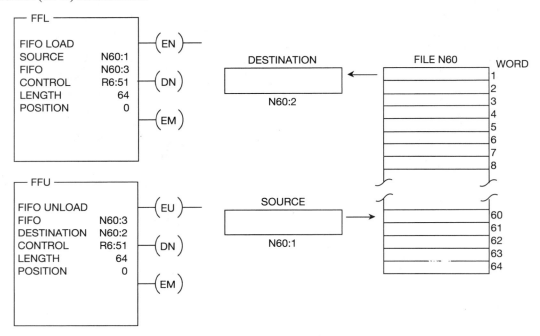

Figure 16–22 PLC-5 FIFO Load and Unload Instructions

The source is the address that stores the "next in" value to the stack. The LOAD instruction retrieves the value of this address and "loads" it into the next word in the file, or stack.

Destination is the address that stores the value that exits the file, or stack.

FIFO is the address of the stack. The same address is used for both the FFL and FFU instructions.

Control is the address of the control structure (48 bits—three 16-bit words) in the control area (R) of memory. The control structure stores the instruction's status bits, stack length, and next available position (pointer) in the stack.

When the rung that contains the FFL instruction goes true, the processor sets the EN bit (bit 15) *ON*, and loads the source data (N60:1) into the next available position in the stack. The processor loads data from the source into the stack with each false-to-true transition. When the stack is full, the processor sets the DN bit (bit 13) to 1. The program should be programmed so that when the stack is full, no additional data can be loaded from the source.

When the rung that contains the FFU instruction goes from false to true, the processor sets the EU (enable unload bit 14) to 1, and unloads data from the first element of the stack into the destination word N60:2. As the data is shifted out, the processor shifts the remaining data in the stack *up* one position toward the first word. The processor continues to unload the stack each time the rung goes from false to true until it empties the FIFO stack.

The Gould 984 also uses two instructions for loading and unloading a first in–first out file, stack, table, or queue. Queue means to line up, or wait in line, and is used by Gould in its program manual. Figure 16–23 shows the Gould 984 first in (FIN) instruction.

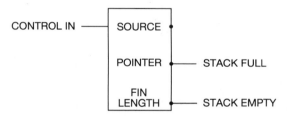

Figure 16–23 984 First In (FIN) Function Block

The top node is the address of the data source. The middle node *must* be a holding register, and acts as the pointer to the destination table, or stack. The bottom node contains the symbol (mnemonic) for first in (FIN), and also holds the stack length. The maximum stack length is 100.

The Control In line moves data from the Source into the destination stack on each scan that the instruction is true. A transitional contact is used if single operation is desired. The transitional contact data is only transferred on the first scan on each *OFF* to *ON* transition.

The middle line on the right is the Stack Full line, and is true when the stack is *full*. The bottom line is the Stack Empty line, and is only true when the stack is *empty*. Once data is moved into the stack, this line is set to 0.

Figure 16–24 shows how the FIN instruction operates.

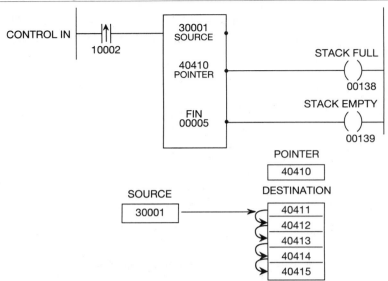

Figure 16–24 Gould FIN (First In) Operation

When the positive transitional contact makes a false-to-true transition, the data in the source word (30001) is transferred into word 40411 in the stack and the pointer is incremented to 1. On the next false-to-true transition, the data in word 40411 is shifted down to word 40412, and the data in word 30001 is copied into word 40411. On each subsequent transition of input device 10001, data from word 30001 is transferred into word 40411, and all previous data is shifted down. The transfer continues on each transition until the stack is full. The first out (FOUT) function is used to unload the stack. As the name implies, the data that was first entered into the stack, or queue, is retrieved first. Figure 16–25 shows how the FOUT instruction operates.

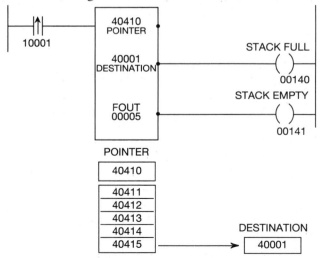

Figure 16–25 984 First Out (FOUT) Operation

On the first transition of input device 10002, the data in word 40415 is copied into the destination register 40001. Now that the stack is no longer full, the stack full line goes false, and output 00140 is set to 0. The processor continues to unload the stack on each transition of 10002 until the stack is empty. When the stack is empty, output 00141 is set to 1.

Last In–First Out (LIFO)

The last in–first out (LIFO) stack is the opposite of the FIFO stack. Information is placed in the stack and then removed in inverse order (the last data in is the first data out). Allen-Bradley uses LIFO load (LFL) and LIFO unload (LFU) instructions for loading and unloading LIFO stacks, or files. The instructions are programmed like the LIFO instructions, and will not be discussed further. The various PLC manufacturers' literature can be consulted if additional information and/or clarification is needed.

Chapter Summary

Although the keystrokes and instructions vary with each PLC manufacturer, the principles of word and file moves are the same. The synchronous shift register shifts bits of information left or right (forward and reverse) within a word or words, while file moves transfer data from words to files, files to words, or files to files. The asynchronous shift register is referred to as FIFO, or first in–first out, as data transfers or falls to the bottom of the stack and uses the last unused word. The data is retrieved in the order it enters the stack (first in–first out). The file can also be programmed to retrieve the data that was last in to be the first data out. This convention is referred to as a LIFO stack.

REVIEW QUESTIONS

1. Define the term *word* as used in this chapter.
2. Define the term *file* as used in this chapter.
3. T   F   The synchronous shift register shifts data in a forward direction only.
4. In a 1-word shift register, the data is entered at bit:
   a. 1
   b. 2
   c. 4
   d. 8
   e. 16
   f. none of the above
5. In a 1-word shift register, the data is shifted out at bit:
   a. 1
   b. 2
   c. 4
   d. 8
   e. 16
   f. none of the above
6. Define the term *FIFO*.

7. Define the term *LIFO*.
8. Briefly describe the difference between *synchronous* and *asynchronous shift registers*.
9. When using the PLC-5 FFL instruction, which bit is set to 1 when the stack is full?
10. When using the PLC-5 FFU instruction, which bit is set to 1 when the stack is empty?
11. List two other names that could be used for a file.
12. When using a Gould 984 PLC, what is the maximum length of a FIFO stack?
13. What is the purpose of the PLC-5 *COP* instruction?
14. Which of the following group of words could *not* be a file?

    a. 50, 51 52

    b. 50, 51, 52, 53

    c.100, 101, 102, 103

    d. 100, 101, 102, 103, 105

15. Briefly describe the function of a word-to-file move.
16. Briefly describe the function of a file-to-word move.
17. Briefly describe the function of a file-to-file move.
18. Which instruction is also known as a FIFO?

    a. synchronous shift register

    b. word-to-file move

    c. file-to-word move

    d. asynchronous shift register

    e. file-to-file move

19. T   F   When data is transferred into a file using a word-to-file move, the data is entered at the last unused word of the file.

Chapter 17 Sequencers

Objectives

After completing this chapter, you should have the knowledge to
- Describe what a sequencer instruction does.
- Understand the basics of sequencer operation.
- Define the term *mask*.

The **sequencer instruction** transfers information from memory words into output words. Sequencer instructions are typically used to control automatic assembly machines that have consistent and repeatable operations.

A programmed sequencer replaces the mechanical drum sequencer that was used in the past. On the mechanical sequencer, when the drum cylinder rotated, contacts opened and closed mechanically to control output devices. Figure 17–1 shows a mechanical drum cylinder with pegs placed at varying horizontal positions for step 1 of the sequence. When the cylinder rotated, contacts that aligned with the pegs closed, and contacts where no pegs existed remained open. In this example, the presence of a peg should be thought of as a 1, or *ON*, and the absence of a peg as a 0, or *OFF*.

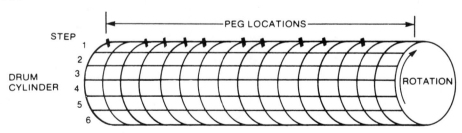

Figure 17–1 Drum Cylinder
(Courtesy of Allen-Bradley)

To program a sequencer, binary information is entered into a series of consecutive memory words. These consecutive memory words are referred to as a **file**. Information from the words in the file is transferred sequentially to the output word to control the outputs.

If the first six steps on the drum cylinder in Figure 17–1 are removed and flattened out, they appear as illustrated in Figure 17–2.

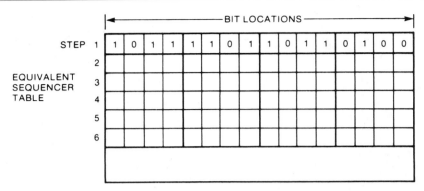

BIT LOCATIONS															
1	0	1	1	1	1	0	1	1	0	1	1	0	1	0	0

Figure 17–2 Sequencer Table

For step 1, each horizontal location where a peg is was located is now represented by a 1 (*ON*), and the positions where there were no pegs are represented by a 0 (*OFF*).

The six steps could also be viewed as a 6-word file with each 16-bit word representing a sequencer step. By entering different binary information (1s and 0s) into each word of the file, the file replaces the rotating drum cylinder.

To illustrate how this works, 16 lamps are used for outputs (as shown in Figure 17–3).

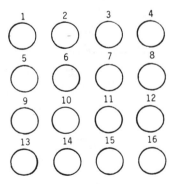

Figure 17–3 Output Lamps

Each lamp represents one bit address (1 through 16) of output word 25.

Assume, for the sake of discussion, that the operator wants to light the lamps in the 4-step sequence shown in Figures 17–4a, 17–4b, 17–4c, and 17–4d.

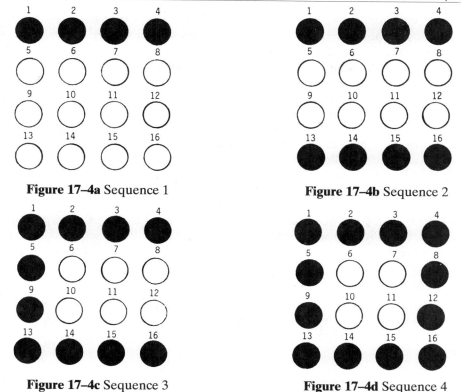

Figure 17–4a Sequence 1

Figure 17–4b Sequence 2

Figure 17–4c Sequence 3

Figure 17–4d Sequence 4

The bit addresses of the lamps that are to be lit in each step of the sequence are written down to assist in entering data into a word file.

The next step is to define a word file to store the binary data required for each step of the sequencer. Words 30, 31, 32, and 33 are used for the 4-word file. By using the programmer, binary information (1s and 0s) are entered into each word of the file to reflect the desired lamp sequence (Figure 17–5).

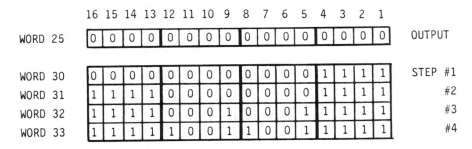

Figure 17–5 Binary Information for each Sequencer Step

NOTE: Some PLCs allow the data to be entered using BCD which speeds the entry process. To use this feature, the required binary information for each sequencer step is converted to BCD. The information is then entered using a programming device into the word file with four key strokes for each word, rather than 16.

Once the sequencer has been programmed and the data entered into the word file, the sequencer is ready to control the lamps. When the sequencer is activated and advanced to step 1, the binary information in word 30 (Figure 17–5) is transferred into word 25, and the lamps light in the pattern shown in Figure 17–4a. Advancing the sequencer to step 2 transfers the data from word 31 into word 25 for the light sequence shown in Figure 17–4b. Step 3 transfers the data from file word 32 into word 25, and step 4 transfers information from word 33 into word 25. When the last step is reached, the sequencer can be reset and sequenced again.

Because each PLC with sequencer capabilities is programmed differently, no attempt will be made to explain the actual programming. Depending on the PLC, sequencers can be programmed from a few steps up to hundreds of steps, and can control one output word or several.

Masks

When a sequencer operates on an entire output word, there may be outputs associated with the word that the operator does not want controlled by the sequencer. To prevent the sequencer from controlling certain bits of an output word, a **mask word** is used. Figure 17–6 shows how a mask word works.

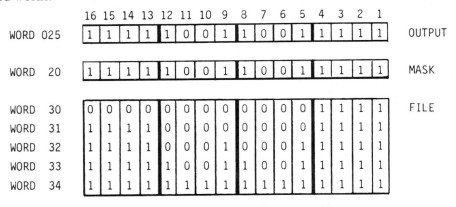

Figure 17–6 Using a Mask Word

The mask word is a means of selectively screening out data from the sequencer word file to the output word. For each bit of output word 025 that the operator wants the sequencer to control, the corresponding bit of mask word 20 must be set to 1.

In Figures 17-4a, 17–4b, 17–4c, and 17–4d, bits 6, 7, 10, and 11 are not used. By *not* setting bits 6, 7, 10, and 11 of the mask word to 1, these bits can be used independently of the sequencer.

In Figure 17–6, a fifth step is added to the sequencer. File word 34 and bits 6, 7, 10, and 11 are set to 1. With bits 6, 7, 10, and 11 of mask word 20 set to 0, the data in file word 34 is screened out and prevented from being transferred into output word 025.

The sequencer works much like the file-to-word move discussed in Chapter 16. For programmable controllers that don't have a dedicated sequencer instruction, a file-to-word move instruction can be used.

Another method of creating a sequencer when no sequencer or file-to-word move is available, is to use a synchronous shift register. A typical synchronous shift register can consist of one or more words. The shift register can shift one bit at a time, 8 bits, or 16 bits (one word) at a time. By using a 16-channel (16-bit shift) multiword shift register, a sequencer can be created.

A 16-channel shift register with five words can also be used to duplicate the sequencer steps shown in Figure 17–5. The synchronous shift register shifts new information in and out on each shift.

To recycle the information and get the cylindrical (repeating) action of a sequencer requires the use of data transfer instructions.

Figure 17–7 shows a 5-word file and a Square D SY/MAX 16-channel synchronous shift register instruction that can be used for the sequencer. Notice that two data transfer instructions are also used.

BITS

	16	15	14	13	12	11	10	9	8	7	6	5	4	3	2	1	WORD	
DATA IN	0	0	0	0	0	0	0	0	0	0	0	0	1	1	1	1	0017	
	1	1	1	1	1	0	0	1	1	0	0	1	1	1	1	1	0018	STEP 4
	1	1	1	1	0	0	0	1	0	0	0	1	1	1	1	1	0019	STEP 3
	1	1	1	1	0	0	0	0	0	0	0	0	1	1	1	1	0020	STEP 2
DATA OUT	0	0	0	0	0	0	0	0	0	0	0	0	1	1	1	1	0021	STEP 1

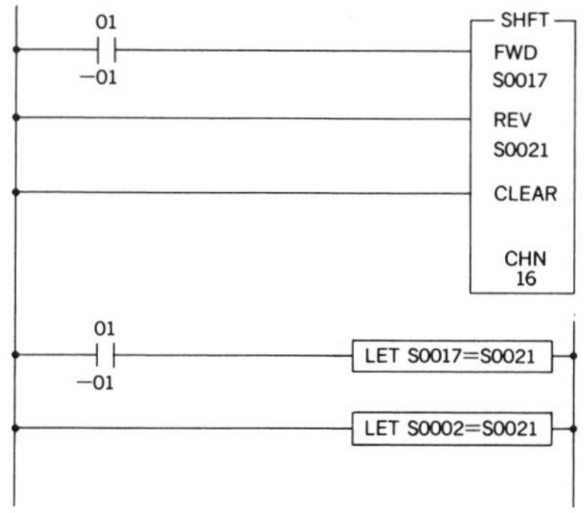
01
–01
SHFT
FWD
S0017
REV
S0021
CLEAR
CHN
16
01
–01
LET S0017=S0021
LET S0002=S0021

Figure 17–7 Programming a 16-Bit Shift Register as a Sequencer

When input 01-01 is closed, the shift register shifts forward. At the same time, the first data transfer rung is also TRUE. The data transfer states that *LET* S0017 = S0021, or tells the processor to take the data in register 0021 and put it in register 0017. This rung has the effect of recycling the data. Normally, when the shift register is shifted forward, the data in register 0021 is shifted out and lost. Now on each shift, the information from register 0021 is transferred *up* and into register 0017. On the next shift, the data in 0017 is shifted *down* to register 0018, and the data from register 0021 is again shifted into register 0017.

The second data transfer rung is programmed unconditional (true all the time) and transfers the data from register (word) 0021 into output word (register) 0002 to obtain the required light display.

The binary information for each sequencer step is entered into each register by using the data mode on the Square D SY/MAX programmer. Although the Square D format was used for this example, the circuit illustrates that knowing and understanding PLC functions allows unique programming possibilities.

Chapter Summary

Although sequencers, like other data manipulation and arithmetic instructions, are programmed differently with each PLC, the concepts are the same. Data is entered into a word file for each sequencer step, and, as the sequencer advances through the steps, binary information is transferred sequentially from the word file to the output word(s). Output word bits can be masked so they can operate independently of the sequencer.

REVIEW QUESTIONS

1. Briefly describe a *sequencer*.
2. A series of consecutive words are referred to as:
   a. deck
   b. group
   c. file
   d. chain
3. What is the purpose of a *mask word* in a sequencer?
4. What device is commonly replaced by a sequencer instruction?

5. Set up the file in Figure 17–A so the sequencer will operate the motors as shown in steps 1, 2, 3, and 4. Program the circuit so motors 01015, 01016, and 01017 cannot be energized.

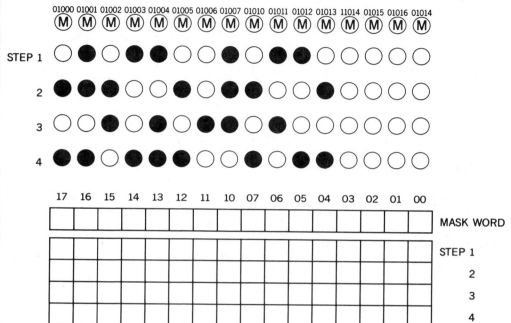

Chapter 18

Programming With Boolean

Objectives

After completing this chapter, you should have the knowledge to
- Define the Boolean term *and*.
- Define the Boolean term *or*.
- Define the Boolean term *not*.
- Interpret truth tables.
- Use Boolean functions to write simple programs.

Boolean Algebra

Don't let the term Boolean or Algebra scare you off at this point. Once the electrician or technician understands some Boolean concepts, he or she will find that programming in Boolean is not only fun, but also fast.

Boolean algebra dates back to 1854 when mathematician George Boole developed his mathematical system of logic. In a logic problem there are only two states: *true* or *false*.

Boolean algebra is a unique system that differs from regular high school and/or college algebra. In the Boolean system there are only two digits: 0 and 1. Every number must be a one (1) or a zero (0), and there are no fractional or decimal numbers, no 2s, 3s, etc.

Relay ladder diagrams can be thought of as logic problems because relay contacts can have only two states: closed (true, or 1), or open (false, or 0).

To better understand the concept, consider the circuit in Figure 18–1.

Figure 18–1 Basic OR Relationship

With contacts A and B wired in parallel, closing either contact A *or* contact B turns on lamp C. This type of circuit arrangement (when either A *or* B can turn on C) is referred to as **OR logic**.

The Boolean equation for an OR statement would be written A + B = C. The plus sign (+) indicates OR logic. According to the formula, if A *or* B is true, then C is true. A truth table for the equation shown in Figure 18–2 further illustrates and clarifies the Boolean concept.

CONTACT	CONTACT	LAMP	
A	B	C	A + B = C
0	0	0	0 + 0 = 0
0	1	1	0 + 1 = 1
1	0	1	1 + 0 = 1
1	1	1	1 + 1 = 1

Figure 18–2 OR Truth Table

NOTE: A 0 represents an open, or *OFF*, condition, whereas a 1 represents a closed, or *ON*, condition.

A logic statement for the series contacts shown in Figure 18–3 shows that if contacts A *and* B are closed, C lights.

Figure 18–3 Basic AND Relationship

This circuit configuration (when both A *and* B must be closed to turn on C) is referred to as **AND logic**. The Boolean formula for an AND statement would be written A × B = C. The multiplication sign (×) indicates AND logic.

NOTE: The AND function can also be expressed as A • B = C or AB = C.

The formula states that if A *and* B are true, then C must be true. A truth table for this equation is illustrated in Figure 18–4.

CONTACT	CONTACT	LAMP	
A	B	C	A X B = C
0	0	0	0 X 0 = 0
0	1	0	0 X 1 = 0
1	0	0	1 X 0 = 0
1	1	1	1 X 1 = 1

Figure 18–4 AND Truth Table

Look at the simple stop/start circuit illustrated in Figure 18–5.

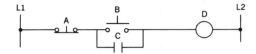

Figure 18–5 Standard Stop/Start Circuit

This circuit contains both AND logic and OR logic. The logic is, if A is closed, *and* either B *or* C is closed, D lights. The Boolean formula is A × (B + C) = D. Figure 18–6 shows the truth table and equations.

CONTACT	CONTACT	CONTACT	LOAD	
A	B	C	D	A X (B + C) = D
0	0	0	0	0 X (0 + 0) = 0
0	0	1	0	0 X (0 + 1) = 0
0	1	0	0	0 X (1 + 0) = 0
0	1	1	0	0 X (1 + 1) = 0
1	0	0	0	1 X (0 + 0) = 0
1	0	1	1	1 X (0 + 1) = 1
1	1	0	1	1 X (1 + 0) = 1
1	1	1	1	1 X (1 + 1) = 1

Figure 18–6 Truth Table for Standard Stop/Start Circuit Logic

Another Boolean concept is the **NOT function.** The NOT function is like the EXAMINE OFF instruction discussed in Chapter 8. The NOT function acts like a set of N.C. contacts. Consider the circuit in Figure 18–7.

Figure 18–7 Basic NOT Relationship

As long as A remains closed (*not* open), lamp B is *ON*. Another approach is to think of the NOT function as being true when the device or address that it represents is *not ON*, or is set to 0. A truth table for a NOT function is shown in Figure 18–8.

CONTACT	LAMP
A	B
0	1
1	0

Figure 18–8 NOT Truth Table

Combining the NOT function with the AND function is illustrated in the circuit and truth table shown in Figure 18–9.

CONTACT	CONTACT	LAMP
A	B	C
0	0	0
0	1	0
1	0	1
1	1	0

Figure 18–9 Circuit and Truth Table for AND and NOT Functions

Combining the NOT function with the OR function is illustrated in the circuit and truth table shown in Figure 18–10.

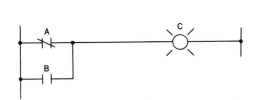

CONTACT	CONTACT	LAMP
A	B	C
0	0	1
0	1	1
1	0	0
1	1	1

Figure 18–10 Circuit and Truth Table for NOT and OR Functions

Combining the AND, OR, and NOT functions are illustrated in the circuit and truth table shown in Figure 18–11.

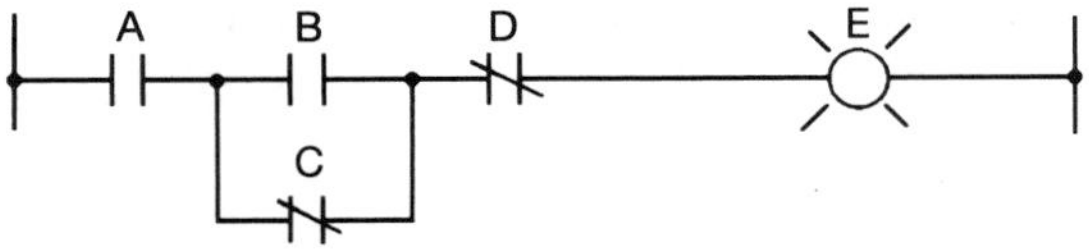

$A \times (B + C) \times D = E$

CONTACT	CONTACT	CONTACT	CONTACT	LAMP
A	B	C	D	E
0	0	0	0	0
0	0	0	1	0
0	0	1	0	0
0	0	1	1	0
0	1	0	0	0
0	1	0	1	0
0	1	1	0	0
0	1	1	1	0
1	0	0	0	1
1	0	0	1	0
1	0	1	0	0
1	0	1	1	0
1	1	0	0	1
1	1	0	1	0
1	1	1	0	1
1	1	1	1	0

Figure 18–11 Circuit and Truth Table Combining AND, OR, and NOT Functions

Programming in Boolean

The techniques used to program a PLC in Boolean are very similar, no matter which manufacturers' PLC is being used. Some of the small PLCs programmed in Boolean are the Gould PC0085, Square D Model 50 and Micro-1™, and the GE-Fanuc Micro. The model or manufacturer is not as important as the concepts that are involved. To program simple control logic with the Gould PC0085, the following seven keys are used:

1. STR   (Store)
2. AND   (AND function)
3. OR   (OR function)
4. NOT   (NOT function)
5. OUT   (Output)
6. ENTER   (ENTER instruction)
7. NEXT ADRS   (Next address)

Figure 18–12 shows a simple line of ladder logic and the keystrokes required for programming.

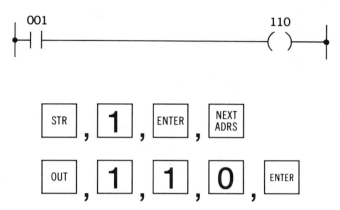

Figure 18–12 Programming Simple Line of Ladder Logic

Programming is started by pushing the STR (Store) key followed by up to a 3-digit address.

NOTE: It is not necessary to enter zeros that proceed numbers in addresses. The address 001 requires that only a 1 be pushed. Similarly, the address 050 requires that only a 5 and a 0 be pushed.

The STR key followed by the number 1 enters a N.O. contact addressed 001.

Pushing the ENTER and NEXT ADRS (next address) keys loads the N.O. contact 001 and readies the processor for the next instruction. Pressing the OUT (output) key and then the 3-digit number 110 gives us the output address 110. Pressing the ENTER key loads the rung of logic.

NOTE: Because the OUTPUT was the last element in the program, only the ENTER key was needed to load the information. If an additional rung (or rungs) was to be programmed, then both the ENTER and NEXT ADRS keys would be used.

Once the programming is complete, the processor is placed in the *monitor* mode (*run* mode) to verify the program.

Figure 18–13 shows two contacts in series (AND function) and the keystrokes required for programming.

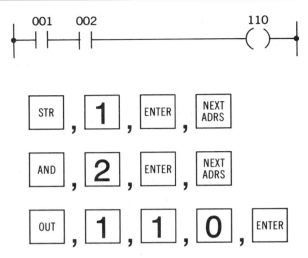

Figure 18–13 Programming AND Function

The keystrokes: STR, 1, gives us an N.O. contact 001. The keystrokes: AND, 2, puts an additional N.O. contact (002) in series with N.O. contact 001. The keystrokes OUT, 1, 1, 0, provides output 110 to complete the line of logic. To program normally closed (N.C.) contacts (or a NOT function) the STR key is followed by the NOT key (shown in Figure 18–14).

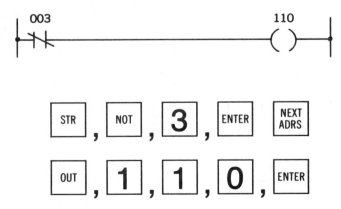

Figure 18–14 Programming NOT Function

Figure 18–15 shows a circuit with an N.O. contact 001 in parallel (or ORed with a N.C. or NOT contact 003) controlling output 110. The keystrokes used to program the circuit are also shown.

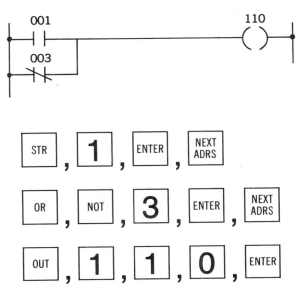

Figure 18–15 Programming OR Function

For more complex circuits where AND and OR functions are combined, the processor uses a temporary file, or stack system, to store information as the circuit is being developed. Each time the STR key is pushed, the information that follows is put into the temporary file or stack. Information enters the stack from the bottom and pushes previous information up. Figure 18–16 illustrates this concept.

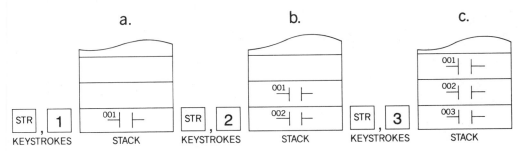

Figure 18–16 Entering Information into Temporary File (Stack)

NOTE: Because it is understood that the ENTER and NEXT ADDRESS keys *must* be used during each step of program development, these keystrokes will no longer be shown.

When the first STR (STR, 1) was entered at (a), N.O. contacts 001 were placed on the bottom of the stack. When the second STR (STR, 2) is entered at (b), N.O. contacts 002 are inserted at the bottom of the stack, and contacts 001 move up one place. At position (c), the third STR command is entered and N.O. contact 003 enters the bottom of the stack. The previous entered contacts again move up. At any point in the program, the contacts can be retrieved in inverse order (last in–first out), or LIFO.

The keystrokes AND, STR take the last contact entered (003) and AND it with the next contact up in the stack (002). Figure 18–17 shows what the stack looks like after the AND, STR commands are entered.

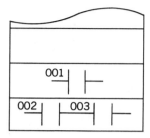

Figure 18–17 LIFO Stack after AND, STR Commands

If the keystrokes OR, STR had been entered, the contacts 003 and 002 would have been ORed as shown in Figure 18–18.

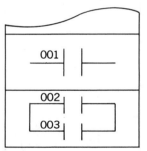

Figure 18–18 LIFO Stack after OR, STR Command

With the contacts in the order shown in Figure 18–18, an AND, STR, series of keystrokes would AND the ORed contacts 002 and 003 with contacts 001, as shown in Figure 18–19.

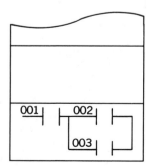

Figure 18–19 LIFO Stack after AND, STR Command

Different STR commands yield different results on the original stack in Figure 18–16. Figures 18–20a, 18–20b, and 18–20c show some alternative keystrokes, and the results of each.

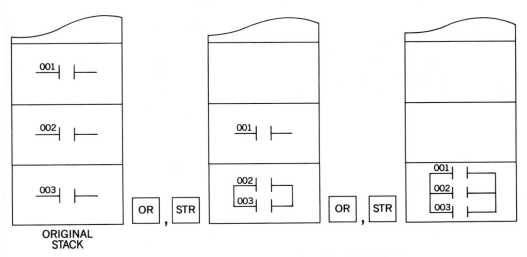

Figure 18–20a Stack after Two Consecutive OR, STR Commands

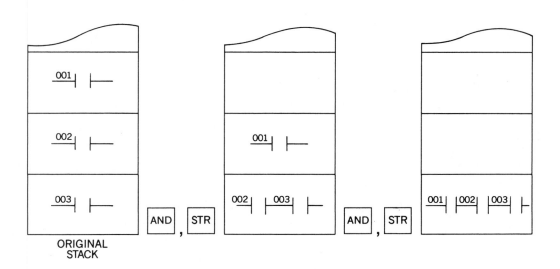

Figure 18–20b Stack after Two Consecutive AND, STR Commands

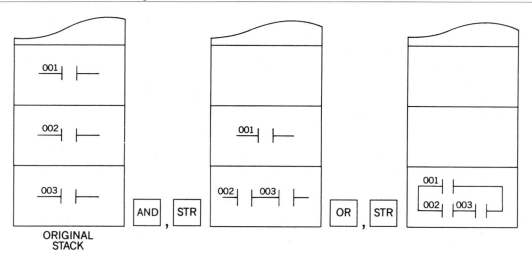

Figure 18–20c Original LIFO Stack after AND, STR and OR, STR Commands

The circuits shown in figures 18–20a, 18–20b, and 18–20c could have been programmed without first developing a stack of three contacts. This method was used to demonstrate a concept, and was used for illustration only.

The keystrokes used to program the three parallel contact shown in Figure 18–20a are:
    STR, 1
    OR, 2
    OR, 3

To program the contacts in series as shown in Figure 18–20b, the following keystrokes are used:
    STR, 1
    AND, 2
    AND, 3

The following keystrokes are used to program the parallel series combination circuit shown in Figure 18–20c:
    STR, 2
    AND, 3
    OR, 1

To complete the circuits shown in these figures, the OUTPUT key is pressed and a 3-digit address entered. Although not all small PLCs allow for parallel outputs, Gould's PC0085 does. Figure 18–21 is an example of parallel outputs and the keystrokes required for programming.

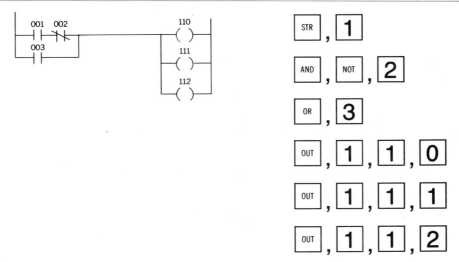

Figure 18–21 Programming Parellel Outputs

NOTE: Remember that ENTER and NEXT ADDRESS follow each step of programming, and are not shown for the sake of simplicity.

As stated earlier in this chapter, programming in Boolean is fun and fast. Notice that no branch *start* or branch *end* commands are needed, which speeds programming time.

To further illustrate the programming techniques and the LIFO (last in–first out) stack concept, Figure 18–22 shows a series parallel combination circuit with the keystrokes and the temporary stack results.

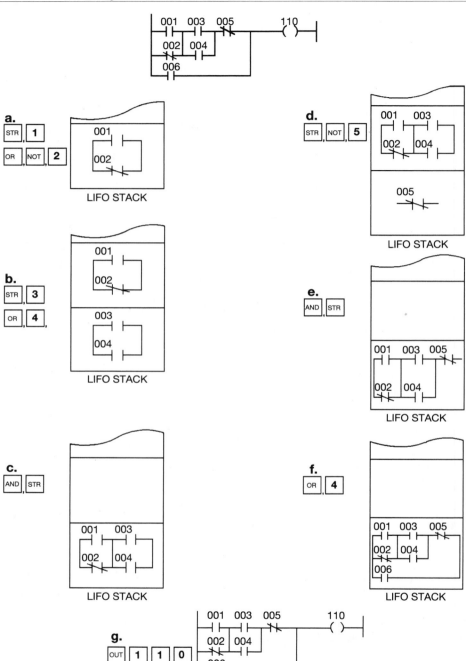

Figure 18–22 Programming a Combination Circuit and the Resulting LIFO Stack

The keystrokes in (a) connect contacts 001 and 002 in parallel, and place them in the LIFO stack.

The keystrokes in (b) connect contacts 003 and 004 in parallel, and enter this data into the stack. This moves contacts 001 and 002 up to make room for contacts 003 and 004 on the bottom.

The next step (c) brings contacts 001 and 002 back down to the bottom of the stack, and ANDs them with contacts 003 and 004.

The STR, NOT, 5, keystrokes at (d) moves the ANDed contacts up in the stack, and a N.C. NOT contact (005) moves into the bottom of the stack.

The AND, STR command at (e) recalls the contacts to the bottom of the stack, and ANDs them with NOT contact 005.

The next step (f) recalls the contacts from the LIFO stack, and ORs them with contact 006.

In the final step, at (g), the output is added with the keystrokes OUT, 110.

Timers

Programming timers with the Gould PC0085 is quick and easy. Figure 18–23 shows a simple on delay timer circuit and the keystrokes necessary for programming.

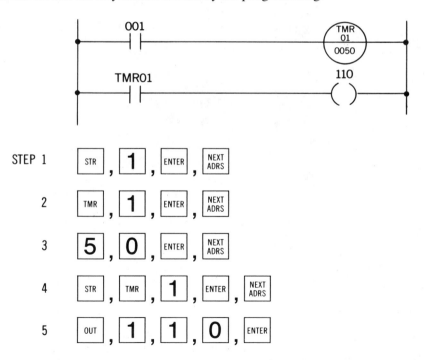

Figure 18–23 Programming Simple ON DELAY Timer

Step 1.   The contact to control the timer is entered (N.O. 001).

Step 2.   A timer is entered and identified as timer number 1.

Step 3.   The preset for timer 1 is entered. The time base for PC0085 timers is 0.1 seconds. Timer 1 in this illustration is 5 seconds (50 × 0.1 seconds = 5 seconds).

Step 4.   A set of timer contacts is entered that will close five seconds after timer 1 (TMR 1) is energized.

Step 5.   The circuit is completed with output 110. Output 110 is controlled by timer contacts, and is delayed by five seconds after input device 001 closes.

The examples and discussion in this chapter provide only an introduction to programming in Boolean, but provides a basic understanding of the concept to enable an electrician or technician to read the user manuals for any small PLC, and to get started programming.

Defining Small PLCs

When we talk about small PLCs, what do we mean by small? "So small it fits in a shoe box" is how one manufacturer describes its product. Figure 18–24 shows a Gould PC0085. (The Number 2 pencil is used to show relative size.) In this case, small is $5^3/_4$"W X 6"H X 4"D.

Figure 18–24 Gould PC0085 Small Programmable Controller
(Courtesy of Gould Inc.)

Depending on whose definition you use, PLCs can be grouped by I/O size.

A PLC users' magazine uses the following I/O sizes to categorize PLCs:
    Micro—up to 64 I/O
    Small—65 to 128 I/O
    Medium—129 to 892 I/O
    Large—more than 892 I/O

The terms **micro**, **small**, **mini** and **compact** are used interchangeably by PLC manufacturers. Although there isn't agreement on what to call this line of miniature PLCs, they all have several things in common: they are small, flexible, modular, and inexpensive. These PLCs are great for replacing existing relays and relay control circuits, and like their big brothers, offer timer and counter functions, math functions, data compare, data transfer, shift registers, sequencers, etc. Although not every small PLC offers all these functions, one can undoubtedly be found to meet specific needs.

A typical small PLC has 800–1000 words of memory. Memory type is CMOS RAM with a lithium battery for backup. Optional storage or backup might be EEPROM or E^{2}PROM. Software is available for many of the small PLCs that allows them to be programmed from an IBM® compatible computer. Once a program has been developed, it can be stored on the hard drive or on a floppy disk.

Chapter Summary

Programmable controllers can be grouped by I/O size. Generally, any PLC with up to 128 I/O is categorized as a small PC. The small PLCs are small physically, but offer many of the features of larger, more expensive, PLCs. Many of the small PLCs are programmed using limited Boolean statements rather than RELAY LADDER LOGIC. Dedicated programmer costs are high (some cost more than the processor, power supply and I/O section combined). Because cost is usually a factor in the small PLC market, Boolean programming is used to help create a system with minimal programmer costs. Typical Boolean functions are AND, OR, and NOT. Programming a small PLC that uses Boolean logic is fast and easy once the basics are understood.

REVIEW QUESTIONS

1. T   F   Boolean algebra was developed in the 1960s when PLCs were first being used.
2. In the Boolean system, which of the following digits are used?
   a. 1, 2, 3, and 4
   b. 1, 3, and 5
   c. 0, 1, 2, and 4
   d. 0, 1, and 2
   e. none of the above
3. What Boolean relationship do contacts have that are connected in parallel?
   a. AND
   b. OR
   c. NOT
   d. IF
4. In a Boolean equation, a plus sign (+) indicates which type of logic?
   a. AND
   b. OR
   c. NOT
   d. IF

5. A multiplication sign (×) indicates what type of Boolean logic?
   a. AND
   b. OR
   c. NOT
   d. IF
6. The NOT function in a circuit acts like a set of:
   a. N.O. contacts
   b. N.C. contacts
   c. timed contacts
   d. intermittent contacts
7. Using the Gould PC0085 programming format, list the keystrokes necessary to program the circuits shown below.

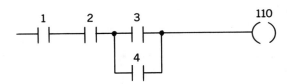

8. List the keystrokes necessary to program the following circuit using a Gould PC0085.

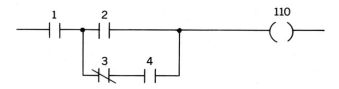

9. What keystrokes are required to enter the following program into a Gould PC0085 memory?

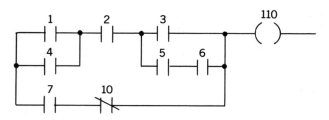

10. Complete a truth table for the circuit shown below.

CONTACT 1	CONTACT 2	CONTACT 3	OUTPUT 110

11. Complete a truth table for the circuit shown in Question 8.

CONTACT 1	CONTACT 2	CONTACT 3	CONTACT 4	OUTPUT 110

12. Define the term *LIFO*.

Chapter 19

Understanding Basic MS-DOS® Commands

Objectives

After completing this chapter, you should have the knowledge to

- Format a new disk
- Create files and directories
- Copy files and directories
- Rename files and directories
- Use the *tree* command

Microsoft Corporation, in cooperation with IBM®, developed software known as **MS-DOS®**, which is an acronym for **MicroSoft Disk Operating System**. The number that follows the word DOS is the version number. For example, DOS 1.0 (version 1) was introduced in 1981, and since that time, there have been many revisions to the original program. The current version of DOS is version 6.1, and as the software is expanded and enhanced, each subsequent version will be assigned a new number (6.2, 6.3, etc.). When the software is upgraded, the software engineers try to maintain **upward compatibility**. That means that the features of the older version are supported, and can be used along with the new features of the revised version. This allows files created in an older version of DOS to be used in a newer version. The examples and commands used in this chapter are for DOS 6.0.

Starting the Computer

DOS software is normally stored on the hard drive of the computer, but when using a computer without a hard drive, the DOS program is stored on a diskette. To use the software, the computer must first be **booted**. This term indicates that the computer has been turned on, and the software loaded into memory. When the computer is first turned on, it is said to be **cold started**. Similar to a PLC, the first thing the computer does is run a self-diagnostic test (referred to as a **power-on self-test** or **POST**). After the initial self-test is run and passed, the computer reads the DOS software from either the diskette or the hard drive. Once the program is read, the **system prompt** is displayed. The prompt is for the **default drive**. On most computers *with* a hard drive, the default (or normal) drive is drive C. Therefore, the prompt that appears is C, followed by a colon (:). On a computer *without* a hard drive, the default drive is usually drive A. Changes from the default drive to another drive are made by typing the letter of the desired drive, followed by a colon. For example, if the default drive is drive C, and a change to drive A is needed, at the prompt, the letter A is typed followed by a colon(:).

EXAMPLE:
Original prompt   C:
Type A:, and hit the RETURN, or ENTER key

The prompt is changed to A:. Drive A is now the primary drive and programs can be loaded or run from that drive.

Formatting a Disk

Once the computer is booted (turned on and operational) and the DOS program loaded, commands can be entered. One of the first commands that should be learned is how to format a disk. Although it is possible to buy formatted disks, the electrician or technician may have to format the disk prior to using it. The formats for IBM compatible computers and Apple® computers, for example, are quite different. Therefore, the discussion here is only for IBM compatible computers running DOS. There are basically two sizes of floppy disks: $5^1/_4$" and $3^1/_2$". Both are available in **double sided, double density (DS,DD)** and **double sided, high density (DS, HD)**. The storage capacity is different for each size and density, and are as follows:

$5^1/_4$" DS,DD 360K
$5^1/_4$" DS,HD 1.2 M

$3^1/_2$" DS,DD 720K
$3^1/_2$" DS,HD 1.44M

The $3^1/_2$" DS,HD disk is the most common disk used today because the most common drive is high density.

NOTE: Storage is rated in K bytes and M bytes. K stands for *kilo* and equals 1,000. M stands for *mega* and equals 1,000,000.

To format a new $3^1/_2$" DS,HD disk, place the disk in the appropriate drive (in this example, drive A is the $3^1/_2$" high density drive. At the C: prompt, type FORMAT A: and hit the ENTER key. The DOS software shows the following prompt:
Insert new diskette for drive A:
and press ENTER when ready. . .

Once the new $3^1/_2$" DS,HD disk is inserted, and the RETURN or ENTER key is pressed, the next prompt reads:
Checking existing disk format.
Saving UNFOMAT information.
Verifying 1.44M

NOTE: The "verifying 1.44M" message indicates that the software is checking the disk to make sure it is 1.44 megabytes. To format a $3^1/_2$" DS,DD disk with 720K of storage in a high density drive, the following command line must be entered at the prompt:
FORMAT A:/F:720

Formatting of the disk now begins, and the percent of the format that is completed is displayed on the screen. Once the disk format is complete, the prompt indicates "Format Complete." The next prompt asks if a **volume label** is to be assigned. The volume label is used to help identify the diskette. The label appears as an entry before the listing of files and helps to verify which diskette is being accessed. The volume label can be up to 11 characters in length. These characters can include letters, numbers, and spaces only: *no* special characters or punctuation can be used. If a volume label is not desired, press the ENTER key to indicate "None." The screen then lists the bytes of total disk space *used*, and the total disk space *available* for use. The prompt then asks if another disk is to be formatted (Y/N)?. If a second disk is to be formatted, remove the previously formatted disk and insert a new one, and press the Y key to initiate the formatting procedure.

Creating Files and Directories

Once a disk has been formatted, it can be used to store new information (PLC programs, files, etc.) or to store existing information as a backup to the hard drive. Before a program can be stored, it must be given a name so it can be located at a later time. The name given to the program is its **file name**. The file name may also have an **extension** to help identify the contents of the file. The original file name cannot be more than eight characters in length, and the extension cannot be more than three characters. Therefore, the file with its extension cannot exceed eleven characters (the period that separates the file name and the extension is *not* counted as a character). If a program is written for a machine that mixes different chemicals to make a lawn fertilizer, the file could be named *lawn*. If the mix uses 20 percent nitrogen, an extension could be added to the file name to further identify the program. An extension is added by placing a period (.) *after* the original file name and then adding the extension. The new file name with the extension in this example would be *lawn*.20%. MS-DOS has rules for naming files. The file name may contain letters, numbers, and some special characters, but for the sake of simplicity and ease in remembering file names, it is normal to use only letters and numbers.

As a general rule, use letters A through Z and the numbers 0 through 9. If desired, the following special characters could be used:

tilde (~)
exclamation point (!)
at sign (@)
number sign (#)
dollar sign ($)
percent sign (%)
caret (^)
ampersand (&)
parentheses ( )
underscore (_)
hyphen (-)
braces ({ })
apostrophe (')
grave accent (`)

None of the other special characters can be used.

File names and extensions cannot contain spaces, commas, backslashes (\), or periods (except when the period is used for the extension to the original file name).

File names and extensions cannot use the following names that have been reserved for use by DOS: AUX, CLOCK$, COM1 through COM4, CON, NUL, and PRN.

Some program software requires that a file name be entered prior to the start of the programming process, whereas other software does not require that a file name be assigned until the program is complete or is to be stored or saved. PLC programs usually do not ask for an extension because the program automatically assigns an extension when the file is saved. However, once a number of files have been created, it becomes necessary to manage them so that any given file can be retrieved on demand. The disk (hard drive or diskette) where the files are stored should be thought of as a file cabinet. If the files are simply thrown into the cabinet, it is difficult to find them at a later date. **Directories** are used to organize the way in which files are stored, and are similar to the drawers in a file cabinet.

If 20 or 30 different files (programs) for various mixes of lawn fertilizer are created, it is easier to organize them if a directory, or drawer in the file cabinet, is named LAWN. The files containing the various *lawn* mixes are then placed into this directory (see Figure 19–1).

Directories, like file names, can also have extensions, although the same restrictions apply to

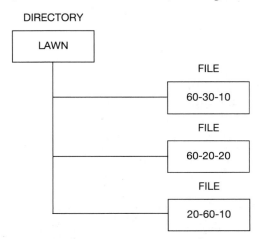

Figure 19–1 Directory with Files

naming directories as those that are applied to naming files. If large numbers of files are under one directory, it may be necessary to create **subdirectories** to eliminate confusion and speed the retrieval process. For example, within the LAWN directory, subdirectories named NITROGEN and POTASH could be created. Each subdirectory would then contain the appropriate files. A block diagram of a directory with subdirectories and files is shown in Figure 19–2.

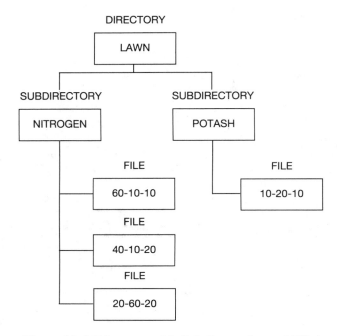

Figure 19–2 Directory with Subdirectories and Files

The directory that contains subdirectories is sometimes called the **parent directory**, and a subdirectory is sometimes called a **child directory**. The subdirectory can also be thought of as a file folder that holds files. The directory is the *drawer* in the file cabinet, and the subdirectory is the *file folder* in the *drawer*. When naming subdirectories, the same rules used for file names and directories must be followed. Subdirectories can also have 3-character extensions.

To retrieve a file, it is necessary to specify the **path** so the computer knows the location of the file. To retrieve file 20-60-20 from subdirectory NITROGEN in directory LAWN on drive C, the following path is entered at the prompt: C:

C:\LAWN\NITROGEN\20-60-20

NOTE: The backslash (\) is used to specify the path from directory to subdirectory to file. Although the examples all appear in capital letters, it is not necessary to enter commands in capital letters.

When additional programs are written, they are given file names as well as paths so they are stored in the appropriate location. For example, if a program is written for a new potash fertilizer mix that is 20-20-10, the following drive path series are entered to place file 20-20-10 into the subdirectory POTASH in the directory LAWN:

C:\LAWN\POTASH\20-20-10

This instructs the computer to save file 20-20-10 on drive C in directory LAWN and subdirectory POTASH. If the file is to be stored on a diskette in drive A, the following drive path is entered at the prompt C:

A:\LAWN\POTASH\20-20-10

If only the file name was entered, it would have been saved in the default, or current directory, of the specified drive. Although the file would have been saved—thrown in the file cabinet—it might be difficult later to remember where it is. Specifying the directory where a file is to be stored makes retrieval easier and faster.

Directory (DIR) Command

DOS keeps track of all the directories, subdirectories, and files that are created. To obtain a list of the directories on the hard drive or on a diskette, the DIR command is used. To look at the directories stored on the hard drive, at the prompt C: type DIR, and hit the ENTER (or RETURN) key. The directory of drive C is displayed. If a large number of directories are present on the drive, they scroll rapidly from the start to the end of the directory listings. Only the last screen full of directories is displayed. To view all the directories a screen full at a time, an additional command is used with the DIR command to pause, or page through directories. To page, or go through the directories one screen full at a time, type a slash (/) and the letter P after the DIR command. The new command at the prompt is:

C: DIR/P

Press the ENTER key, and only one page, or screen, of the directory is displayed. A new prompt appears, stating "Press any key to continue." This allows the operator to view the listing one page (screen full) at a time.

To look at the directories on drive A, type:

C: DIR A:

When the ENTER key is pressed, the directory of drive A is displayed on the screen. If the directory on drive A is too long and fills more than one screen, the page command can be entered after the A:. The command line is:

DIR A:/P

It is also possible to pause the screen by pressing the control (CTRL) key and then the S key. This causes the directory screen to stop scrolling, so the directory listing can be read. Press any key to continue scrolling the listing. A third way to stop the screen from scrolling is to use the PAUSE/BREAK key. This key is normally found in the upper right-hand section of the normal keyboard. Pressing this key pauses the scrolling of the screen. To start the screen scrolling again, press any key to continue. Figure 19–3 shows a typical directory listing.

```
Volume in drive C has no label
Volume serial Number is 2A4C-11DF
Directory of C:\
COMMAND    COM             52925   03-10-93      6:00a
DOS              <DIR>             08-16-93     10:22a
WINDOWS          <DIR>             08-16-93     10:23a
MOUSE            <DIR>             08-16-93     10:25a
PRODPACK         <DIR>             08-16-93     10:25a
TEMP             <DIR>             08-16-93     10:26a
DATA             <DIR>             12-20-93      3:00p
CHKLIST    MS                189   03-29-94      5:23p
SCANV      BAT               688   04-08-93      2:39p
SYSBACK    BAT              7866   04-02-93      5:00p
WINA20     386              9349   03-10-93      6:00a
DOS_DISK         <DIR>             08-16-93     10:29a
WIN_DISK         <DIR>             08-16-93     10:30a
NC               <DIR>             09-27-93      3:27p
AUTOEXEC   BAK                37   03-29-94      5:22p
CONFIG     SYS               232   01-25-94      5:05p
PF               <DIR>             09-27-93      3:50p
WP51             <DIR>             09-27-93      3.35p
123              <DIR>             09-27-93      4:10p
Press any key to continue. . .
```

Figure 19–3 Directory Listings

Directories are indicated by <DIR>. The other listings are files. Notice that there is a date and time after each listing. This indicates the last date and time changes were made to the directory and/or files.

When the directory is too long to be displayed on one screen, it is possible to reformat the directory using the **wide format display option**. The wide format display lists the directories in five columns without dates and times. In many cases, this technique allows all of the directories and files to be displayed. To display the directory listings in wide format as shown in Figure 19–4, at the prompt type:

DIR/W

```
Volume Serial Number is 2A4C-11DF
Directory of C:\
COMMAND.COM      [DOS]          [WINDOWS]       [MOUSE]         [PRODPACK]
[TEMP]           [DATA]         CHKLIST.MS      SCANV.BAT       SYSBACK.BAT
WINA20.386       [DOS_DISK]     [WIN DISK]      [NC]            AUTOEXEC.BAK
CONFIG.SYS       [PF]           [WP51]          [123]           [CLIN]
DIR              [LMODSOFT]     CONFIG.01       GDCP.PAR        [UTIL]
[SPSS]           COXCO]         SPSS.LOG        SPSS.LIS        SCRATCH.PAD
[BOOK]           [BAT]          AUTOEXEC.BAT    [RENTAL]        [RES.PRJ]
[MISC]           [ADV.ELE]      WASTE           [QAPLUS]        [WINFAX]
[CCM]            [TELIX]        [DOWNLOAD]      COMB1.PRG       [LAWN]
[LETTERS]        [IPDS]         AB.BAT          TREEINFO.NCD    [CTC]
                 50 files(s)    123040 bytes
                                                171151360 bytes free
```

Figure 19–4 Wide Display Option

Interrupt Command (Break Command)

The **interrupt command**, or **break command**, is used to interrupt, or stop, the existing command and return to the prompt. The command consists of the CTRL key and the C key. For example, if only a portion of the directory is to be viewed, press the CTRL key, and while holding it down press the C key. This ends the directory listing and returns to the prompt. This combination of keys always interrupts whatever command is entered and returns to the prompt. If the computer has been configured for BREAK, pressing the CTRL key and the PAUSE/BREAK key also interrupts the current command and returns to the prompt. These commands are also called **break commands**.

Clear Screen Command

After entering several commands, the screen becomes crowded. To clear the screen, the **clear screen command** (**CLS**) is used. At the prompt type:

CLS, and press the ENTER key

The screen is cleared of all previous commands and displays, and only the prompt appears at the top left of the screen.

Copy Command

The **copy command** is used to copy files from one disk to another, from one disk to the hard drive, or from the hard drive to a diskette. To copy file 10-20-10 from a diskette in drive A to a diskette in drive B, the following command line is typed at the C: prompt:

COPY A:10-20-10 B:

When the ENTER key is pressed, file 10-20-10 is copied from the diskette in drive A to a diskette in drive B. In this example, drive A is the **source drive** and drive B is the **target drive**. If no source drive is given, DOS assumes that the file is on the default drive. In the previous example, drive C was the default drive.

If a file named TIMERS from the hard drive (drive C) is to be copied to a diskette in drive A, the following command line is used at the C: prompt:

COPY TIMERS A:

When the ENTER key is pressed, the file TIMERS is copied from drive C to a diskette in drive A. A source drive is not designated because the file is on the default drive, and DOS assumes the source drive to be the same as the default drive.

To save time in copying all the files on one disk to another, the copy command is followed by and asterisk (*), a period (.), and another asterisk (*). To copy all the files on a diskette in drive A onto a diskette in drive B, the following command line is entered at the C: prompt:

COPY A:*.* B:

All the files on the diskette in drive A are copied onto the disk in drive B. Because no file names were specified, the file names from diskette A are used for the file names on diskette B.

The partial directory shown in Figure 19–3 can be used to further illustrate the copy command. To copy the COMMAND.COM file from drive C to a diskette in drive A, the following copy command is used at the C: prompt.

COPY COMMAND.COM A:

When the ENTER key is pressed, file COMMAND.COM on drive C is copied onto the diskette in drive A. In this example, the C drive is the source, and the A drive is the destination.

To copy the file in this example, but change the name of the file to COMM.2 on drive A, the following copy command is entered at the C: prompt.

COPY COMMAND.COM A:COMM.2

When the ENTER key is pressed, the COMMAND.COM file is copied from drive C to drive A, and the name is changed to COMM.2.

To copy a file that is in a directory and/or subdirectory, the drive path must be specified. In Figure 19–2, the mix for a certain fertilizer (10-20-10) was stored in directory LAWN and subdirectory POTASH. To copy file 10-20-10 onto a diskette in drive A, the following copy command is typed at the C: prompt.

COPY LAWN\POTASH\10-20-10 A:

The drive path specified tells the computer to copy file 10-20-10 from directory LAWN and subdirectory POTASH from C drive to A drive. The copy command cannot be successfully executed unless the full drive path is specified.

Making Directories Command

The **make directory command (MD)** is used to make a directory on a diskette or on the hard drive. At the prompt, type MD, and then the name of the directory. If a directory named PLC-5 is to be created, the following command is entered at the C: prompt:

MD PLC-5

When the ENTER key is pressed, a directory named PLC-5 is created on drive C. To verify that the directory was created, use the DIR command. At the C: prompt type:
  DIR PLC-5

When the ENTER key is pressed, the directory named PLC-5 is displayed. If the directory was not created, the screen reads **file not found**.

To make a subdirctory of an existing directory, both the name of the directory and the subdirectory need to be entered. To create subdirectory NITROGEN under directory LAWN, type the following command line at the C: prompt:
  MD LAWN\NITROGEN

When the ENTER key is pressed, a LAWN directory with a subdirectory NITROGEN is created. Notice that the directory name must be entered first, followed by the subdirectory name (names).

Change Directories Command

To change from one directory to another, the **change directory command (CD)** is used. To move from the LAWN directory to the PLC-5 directory, type the following command at the C: prompt:
  CD\PLC-5

When the ENTER key is pressed, the software changes from the LAWN directory to the PLC-5 directory.

Renaming Files Command

The **renaming files command (REN)** is used when it is necessary to rename an existing file. This is done to further clarify the contents of the file, or to avoid duplication. If a new file is given the same name as an existing file, DOS destroys the previous file and creates a new one with the same name. To rename file 10-20-10 to FERT.A, the following command line is typed at the C: prompt:
  REN 10-20-10 FERT.A

This command renames, or changes, the file from 10-20-10 to FERT.A.

Delete Files Command

When it is necessary to delete or erase old files to make room for new files, or to simply delete files that are no longer needed, the **delete command (DEL)** is used. To delete the FERT.A file that was renamed in the previous example, the following command is entered at the C: prompt:
  DEL FERT.A

When the ENTER key is pressed, the FERT.A file is deleted.

Remove Directory Command

To remove an existing directory, the **remove directory command** (**RD**) is used. To remove the PLC-5 directory created earlier, type the following command line at the C: prompt:

RD PLC-5

Directories cannot be removed or deleted until all the subdirectories and files have been deleted.

Directory Tree Command

It is not uncommon to forget what directories, subdirectories, or files are on a given disk or hard drive. The **tree command** is used to provide a list of all directories and subdirectories, and to display their organization. The tree is similar to a family tree in structure. To list the directories and their subdirectories, type the following command at the C: prompt:

TREE

When the ENTER key is pressed, the computer displays all directories and subdirectories. Figure 19–5 shows the results of the tree command.

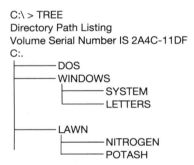

Figure 19–5 Directory and Subdirectory Tree

To list the directories, subdirectories, and files, a /F is added after the tree command and is typed at the C: prompt:

TREE /F

This command produces the tree shown in Figure 19–6.

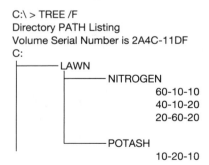

Figure 19–6 Tree with Directories, Subdirectories, and Files

The tree command usually produces more information than can be shown on one screen, and causes the screen to scroll. To stop the scrolling, use the PAUSE/BREAK key.

To view only one directory and its subdirectories, a different command line is used. To view the LAWN directory and its subdirectories, type the following command line at the C: prompt:
TREE LAWN

When the ENTER key is pressed, the tree for the LAWN directory is displayed as shown in Figure 19–7.

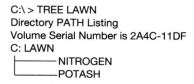

```
C:\ > TREE LAWN
Directory PATH Listing
Volume Serial Number is 2A4C-11DF
C: LAWN
        ├────────NITROGEN
        └────────POTASH
```

Figure 19–7 Tree for LAWN Directory

To look at the directory, its subdirectories, and its files, type the following command line at the C: prompt:
TREE LAWN/F

The results of this command are shown in Figure 19–8.

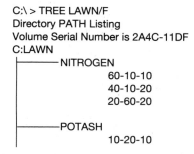

```
C:\ > TREE LAWN/F
Directory PATH Listing
Volume Serial Number is 2A4C-11DF
C:LAWN
        ├────────NITROGEN
                        60-10-10
                        40-10-20
                        20-60-20

        ├────────POTASH
                        10-20-10
```

Figure 19–8 Tree Command for Directory, Subdirectory and Files

To print a copy of the tree, the **print command** (**PRN**) is used in conjunction with a greater than (>) symbol. To print a copy of the tree on drive C, type the following command line at the C: prompt:
TREE > PRN

It is not the intention of this chapter to provide the reader with all of the possible DOS commands, but instead to introduce him or her to the basic commands that are needed to get started using DOS and PLC software packages. Additional information about DOS can be obtained from the many books and software programs that teach the concepts and application of DOS commands. These software programs are referred to as **tutorials**, and can be bought from any retailer that carries computers and software packages.

Chapter Summary

It is common practice to make backup copies of original software disks, to use the copies as working disks, and to make copies of files from the hard drive to a diskette to serve as a backup. It may also be necessary to copy files from one diskette to the hard drive, or to another diskette. Various DOS commands are used to rename files and directories, or to copy files and directories. When using directories and subdirectories, it is important to define the path so the computer will know where to find the file, subdirectory, or directory.

REVIEW QUESTIONS

1. What does the acronym *DOS* stand for?
2. What does *upward compatibility* mean in the context of this chapter?
3. Define the term *booted*.
4. A subdirectory is sometimes called a _____.
5. The directory that contains subdirectories is sometimes called the _____.
6. What command is given at the prompt to look at the directories on the current drive?
7. What command is given to show the directories in the wide version?
8. What does the *break command* do?
9. What command is used to clear the screen?
10. Write the command line that is necessary to copy a file named DEMO from drive A to drive B. Assume the current drive is drive C.
11. Write the command line that is necessary to change the name of the file DEMO to TEST.
12. What is the function of the *tree command*?

Chapter 20

Start Up and Troubleshooting

Objectives

After completing this chapter, you should have the knowledge to
- Understand start-up procedures listed in the manufacturers' literature.
- Explain how input devices are tested.
- Explain how to test output devices using a push button or other input device.
- Explain safety considerations when testing output devices.
- Describe how voltage readings are taken to check input and output modules.

Start Up

Careful start-up procedures are necessary to prevent damage to the driven equipment and the programmable controller system, or, more importantly, injury to personnel.

Prior to beginning a system start-up procedure, it is important to check and verify that the system has been installed according to the manufacturers' specifications, and that the installation meets local, state, and national codes. Special attention should be given to system grounding.

Before applying power to the controller, complete the following steps:

Step 1. Verify that the incoming power matches the jumper selected voltage setting of the power supply. Figure 20–1 shows a typical AC power terminal strip with the jumper position indicated at the sides of the strip.

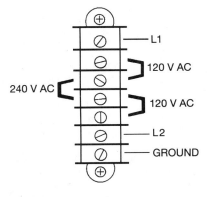

Figure 20–1 Power Supply Terminal Strip with Jumper Positions Indicated

NOTE: For 120 V AC, the neutral conductor is connected to the L2 terminal, and the 240 V jumper is removed.

Step 2.   Verify that a hard-wired safety circuit or other redundant EMERGENCY STOP device (described in chapters 2 and 11) has been installed and is in the open position.

Step 3.   Check all power and communication cables to ensure that connector pins are straight, and not bent or pulled out.

Step 4.   Connect all cables making sure that connectors are fully inserted into their sockets. Secure connectors as applicable.

Step 5.   Ensure that all modules are securely held in the I/O rack, and that field wiring arms (if applicable) are fully seated and locked.

Step 6.   Place the PLC processor key switch to a safe position as indicated in Figure 20–2.

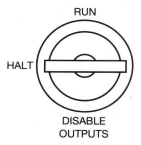

Figure 20–2 Key Switch in Halt Position

Step 7.   Double check the setting(s) of all DIP switches.

CAUTION: Before proceeding further, make sure that the safety circuit or other EMER-GENCY STOP devices are *OFF*, or open, and that power is removed from all output devices.

Apply power and observe processor indicator light(s) for proper indications.

When power is applied and the safety switch is closed, the power supply should provide the necessary DC voltage for the processor and I/O rack. If the proper voltage is present, the input indicator LEDs of the input modules will be functioning. Any input device that is closed, or *ON*, will have an illuminated LED (Figure 20–3).

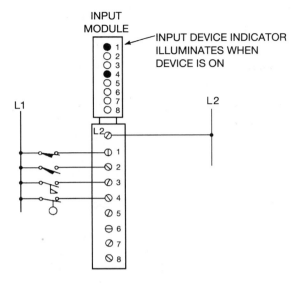

Figure 20–3 Input Module Indicators

Testing Inputs

Each input device can be manipulated to obtain open and closed contact conditions.

> **CAUTION:**   *Do not* activate the input devices mounted on equipment by hand, because unexpected machine motion could cause injury. Use a wooden stick or other nonconducting material to activate input devices mounted on equipment.

Each time an input device is closed, the corresponding LED on the input module should illuminate. Failure of a LED to illuminate indicates:

1. Improper input device operation.
2. Incomplete or incorrect wiring.
   a. Check to be sure that the input device is wired to the correct input module and proper terminal.
3. Loss of power to the input device.
4. Defective LED and/or input module.

To further check the system, a program can be developed and entered into the processor that uses each input device address. With the key switch in the test, or disable output position, the status of the input devices may now be monitored by using the VDT of a desktop or computer programmer, or by LED indicators on a hand-held programmer. On a VDT, the input contact becomes intensified or goes to reverse video, depending on the model PLC, when the instruction is true. An EXAMINE ON instruction intensifies or goes to reverse video when the input device was ON, or closed. An EXAMINE OFF instruction intensifies or shows reverse video when the input is OFF, or open. Figure 20–4 shows a Rung for testing input devices.

DISCRETE INPUT ADDRESS

INTERNAL or DUMMY OUTPUT ADDRESS

11105

02000

Figure 20–4 Rung for Testing Input Devices

Once all input devices have been tested and checked out as operational and properly terminated, the output devices can be tested.

Testing Outputs

Before testing output devices, it must be determined which devices can safely be activated and which devices should be disconnected from the power source. Figure 20–5 shows a motor starter with the motor disconnected for safety. In this configuration, the motor starter coil (the output device) is activated for checkout without energizing the motor. This prevents unwanted machine motion that might cause injury to personnel or damage the machine.

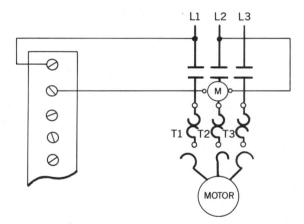

Figure 20–5 Disconnecting Motor Leads for Safety

For outputs that can be safely started, be sure equipment is in the start-up position, is properly lubricated, and is ready to run.

There are two methods used to test output devices. The first method uses a push button or other convenient input device that is part of the control panel. The push button is programmed to energize each output, one at a time.

The second method uses the FORCE function of the PLC to energize outputs, one at a time.

When using a push button (or other input device), the address of the push button is programmed in series with the output device to be tested (shown in Figure 20–6).

ADDRESS OF A N.O.
PUSH BUTTON

ADDRESS OF OUTPUT
DEVICE TO BE TESTED

11105   01000

Figure 20–6 Rung for Testing Output Devices

Once the Rung has been programmed and entered into processor memory, the processor is placed in the *run* mode. Pressing the push button illuminates the output indicator on the output module for address 01000. If there is an output connected, verify that the output device is energized. If the output indicator does not illuminate, using the VDT, verify that the input instruction 11105 is intensified or showing reverse video to indicate an *ON* condition. Double check the output address, and verify that the output instruction indicates an *ON* condition. If both instructions indicate an *ON* condition and the output address is correct, a defective module is likely the problem.

If the output module indicating LED is illuminated, but a connected output device does *not* energize, check the following:
1. *Wiring* to the output device.
2. *Operation* of the output device.
3. Proper *potential* to the output device.
4. Output device wired to correct output module and proper terminal.

A second method for testing output devices (for PLCs with the option) is the FORCE feature. The FORCE feature allows the user to turn an output device *ON* and *OFF* without using a push button or other contacts. Figure 20–7 shows the Allen-Bradley format for programming a Rung to test output address 01000 using the FORCE ON function.

BRANCH END
INSTRUCTION

OUTPUT ADDRESS
TO BE TESTED

01000

Figure 20–7 Rung for Testing Output Devices using FORCE ON Function

The BRANCH END instruction is used to create an open condition in the Rung. This prevents the output from being unconditionally energized when the PLC is placed in the *run* mode. The ALWAYS FALSE instruction that was discussed in Chapter 11 can be used when testing and troubleshooting the PLC-5 family of processors.

NOTE: For other PLC systems, an open instruction would be used to open the Rung.

By moving the cursor to the output instruction, the FORCE ON function can be initiated, which should illuminate the output module indicating LED, and turn *ON* the output device, if one is connected. If the LED does not illuminate, and the output instruction is intensified, exchange the module. If the LED lights, but the output device does not energize, proceed as previously described.

Final System Checkout

After all input and output circuits have been tested and verified, the electrician or technician is ready for the final system checkout.

Reconnect any output loads (motors, solenoids, etc.) that were previously disconnected. In the case of motors, correct rotation needs to be established before the complete machine or process can be tested. Using a momentary push button that is part of the control panel (or using one installed specifically for this purpose), load a Rung of logic into the processor as previously discussed for testing outputs (Figure 20–6).

> **CAUTION:**   Because this part of the test causes machine motion, make sure the machine is operational and all personnel are in the CLEAR. Station someone at the EMERGENCY STOP or disconnect location to deenergize the system, if necessary.

Close the push button and immediately release or open it. This momentary operation of the push button is called **jogging** or **bumping**, and allows the output (motor starter) to energize only momentarily. The motor starter is only energized long enough to determine the direction of rotation of the motor. If the rotation is wrong, reverse any two motor leads and repeat the test for verification. Continue testing all output loads previously disconnected until all of them function correctly. Once all machine components are tested and correct rotation(s) are established, total machine operation testing can be accomplished.

For final system checkout, the following steps should apply:

Step 1.   Place the processor in the *program* mode.
Step 2.   Clear the memory of any previous Rungs used for testing.
Step 3.   Using a programming device, enter the program (ladder diagram) into memory.
Step 4.   Place the processor in the *test* or *disable output* mode, depending on the PLC, and verify correctness of program.

> **NOTE:**   In the *test*, or *disable output* mode, the outputs cannot be energized. All logic of the circuit is verified, input devices function, but no outputs come on. *This step must not be skipped if injury to personnel or damage to equipment is to be avoided.*

Step 5.   Once the circuit operation has been verified in the *test* or *disable output* mode, the processor can be placed in the *run* mode for final verification.
Step 6.   Make changes to the program as required (timer settings, counter presets, and the like).
Step 7.   Once the circuit is in final form, and the machine or process is running correctly, it is recommended that a copy of the program be made.

Troubleshooting

The key word to effective troubleshooting is *systematic*. To be a successful troubleshooter, the technician must use a *systematic* approach.

A systematic approach should consist of the following steps:

Step 1. Symptom recognition.
Step 2. Isolate the problem.
Step 3. Corrective action.

The electrician and/or technician should be aware of how the system normally functions if he or she expects to successfully troubleshoot the system. When prior knowledge of system operation is not possible, the next best source of information, if applicable, is the operator. Don't hesitate to ask the operator what the symptoms are and what he or she thinks the problem might be. If no operator is available, the next best source of information is the PLC system itself. Although the PLC can't talk, it can communicate in various ways to show what the problem is. There are status lights on the processor, power supply, and I/O rack that indicate proper operation, as well as status lights that alert the troubleshooter to the problem. The status lights of a typical processor with built-in power supply indicate:

1. DC POWER ON—If this LED is not lit, there is a fault in the DC power supply. Check the power supply fuse and/or incoming power.
2. MODE—Indicates which operating mode the processor is in (*run*, *halt*, *test*, *program*, etc.). The fault may simply be that the key switch is in the wrong position.
3. PROCESSOR FAULT—When this status light is on, it indicates a fault within the processor. This is a major fault, and requires changing the processor module.
4. MEMORY FAULT—This status light illuminates when a parity error exists in the transmission of data between the processor module and the memory module. Replace only one module at a time. If the first module does not correct the problem, reinstall the original module and replace the second module. If replacing the second module doesn't clear the problem, replace both modules.
5. I/O FAULT—This light indicates a communication error between the processor and the I/O rack. Check that the communication cable(s) are fully inserted into their sockets. If available, connect a programming device with a VDT to the processor, and look for error codes and/or fault messages for further diagnostic assistance.
6. STANDBY BATTERY LOW—When this LED is illuminated, the RAM backup batteries are low and need to be replaced. Although this is not a fault condition, failure to replace batteries results in losing the program when the system is shut down or a power failure occurs.

NOTE: Refer to manufacturers' literature for explanation of error codes.

Status lights on the I/O modules also assist with troubleshooting problems that involve input and output devices.

If the operator confirms that the solenoid that activates a brake isn't working, the first step is to determine the address of the solenoid. Once the address is known, the programming device can be used to ensure that the output circuit to the solenoid has been turned *ON*. Are all input devices closed that should be closed? Has the rest of the Rung logic been completed? If the answers are yes, then it is necessary to determine the hardware location. (Some PLCs use the address to specify the hardware location, whereas others do not.)

Output modules have LED indicators that illuminate when each of the output circuits are turned *ON*. If the LED is lit for the location of the solenoid, it indicates that the problem is not with the output module, but with the circuit from the module to the solenoid, or with the solenoid itself.

A voltage check from L2 to the terminal as shown in Figure 20–8 verifies the conclusion that the module is working properly, but that the problem is either in the wiring to the solenoid, or in the solenoid itself. Further voltage checks will locate the problem.

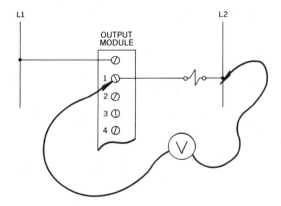

Figure 20–8 Testing Voltage on an Output Module

NOTE: On AC output modules, the high internal resistance of most analog or digital meters act like a series voltage divider when measuring across an open load (Figure 20–9). The result is a reading of nearly full voltage even after the triac has been turned *OFF*. For accurate readings, a 10K ohm resistor can be placed in parallel with the meter leads as shown in Figure 20–10, or a solenoid-type tester (Wiggins) with low internal resistance can be used.

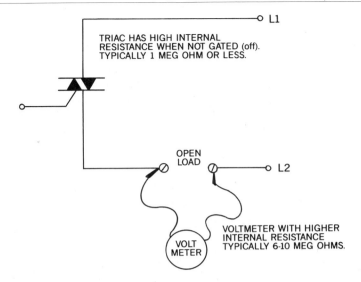

Figure 20–9 Series Voltage Divider Effect when Reading Across an Open Load

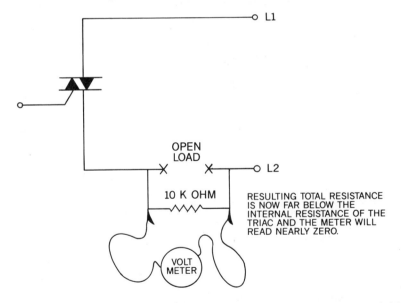

Figure 20–10 Adding a Resistor in Parallel with Meter to Reduce Voltage Divider Effect.

If the indicator LED had not been lit, the first reaction might be to change the output module. Instead, look first for a blown-fuse indicator LED. A fuse might be all that's needed to make the solenoid operational. If, however, there was no blown fuse, replacing the output module should correct the problem.

NOTE: Remove all power from the I/O rack before changing modules.

Some PLC manufacturers, like the Square D Company, offer deluxe output modules that have two indicator LEDs. One indicates that the logic from the processor has been received to turn on the output; the second LED comes ON when the triac, or power transistor, has been turned *ON*. These two LEDs should come on simultaneously, unless the PLC is in the *test* mode. In the *test* mode, only the logic LED is lit because the output circuits are isolated, and kept from being turned *ON*.

Troubleshooting input modules follows the same basic procedure. If it was determined that the solenoid had not energized because limit switch 1 was not being shown closed on the programming device, the LS-1 address would need to be determined. From the address, determine the terminal on the input module that LS-1 is connected to.

An illuminated LED indicates that the limit switch is closed, but that the state of the switch (*ON*) is not being communicated to the processor. Exchanging the input module should make the system operational. However, had the indicator LED for LS-1 not been lit, there could be several other possible problems such as a bad limit switch, faulty wiring from LS-1 to the input module, or a bad input module. Closing the limit switch and taking a voltage check as shown in Figure 20–11 determines if the limit switch and associated wiring is operational.

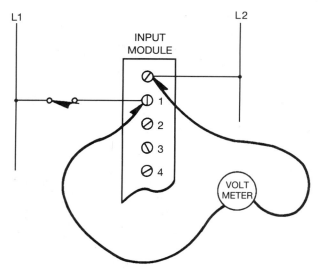

Figure 20–11 Testing Voltage on an Input Module

A voltage reading equal to the applied voltage indicates that the limit switch and wiring are operating, but that there is a faulty input module. No voltage reading indicates a problem with either the limit switch or its wiring. Further voltage checks will isolate the problem.

Similar to the deluxe output modules, there are also input modules that have two indicating LEDs. The first LED indicates that the input device has closed, and a voltage signal has been received by the input module; the second LED indicates that the status of the input device (*ON*) has been communicated to the processor.

NOTE: Once the PLC program has been proven, or checked out, for complete and accurate operation, any future problems normally will be bad field wiring and/or bad field devices.

Chapter Summary

Start-up procedures check each part of the PLC system for proper installation and operation. Safety must *always* be the overriding factor when testing or operating a PLC system. Care must be taken to prevent unexpected or incorrect machine motion if injury to personnel and/or damage to equipment is to be avoided.

Once the system start up is complete and the system is operational, problems or faults can occur. To successfully troubleshoot the system, a systematic approach must be used. This systematic approach includes recognizing the symptoms, isolating the problem, and taking corrective action. A variety of indicator and status lights, as well as error messages and fault codes, assist the electrician or technician in troubleshooting a given system.

The information covered in this chapter is intended to be general in nature and is not specific for any particular PLC. For more specific information on start-up and troubleshooting procedures, refer to the manufacturer's operating manual that accompanies the PLC.

REVIEW QUESTIONS

1. A programmable controller system should be installed according to:
   a. manufacturer's specifications
   b. local electrical codes
   c. state electrical codes
   d. national electrical codes
   e. all of the above
2. Explain why a safety circuit or other EMERGENCY STOP devices are important.
3. Describe briefly how input devices are tested.
4. List two methods for testing output devices.
5. Why would output devices be disconnected before testing?
6. Draw an input module complete with input devices and power connected (L1-L2), and show how a voltage reading is taken to verify the operation of an input device.
7. On a deluxe input module, what do the two LED indicators represent?
8. On a deluxe output module, what do the two LED indicators, other than the blown-fuse indicator, represent?
9. Define the term *jogging*.
10. List the three steps of systematic troubleshooting.

GLOSSARY

ADDRESS: A location in processor memory.

ANALOG INPUT MODULE: A module that converts an analog input signal to a binary or BCD number for use by the processor.

ANALOG OUTPUT MODULE: A module that provides an output proportional to a binary or a BCD number provided to the module by the processor.

ANALOG SIGNAL: A continuous signal that depends directly on magnitude (voltage or current) to represent some condition. For example, a voltage might represent the speed of a motor (5 V corresponding to 200 rpm; 10 V corresponding to 400 rpm, etc.).

AND LOGIC: Logic that has a series relationship. Two devices that are in series would both have to be true to pass the logic. Device one AND two need to be true for the logic to pass.

ARITHMETIC CAPABILITY: The ability of a PLC to perform addition, subtraction, multiplication, division, and other math functions.

ASCII: Acronym for **A**merican **S**tandard **C**ode for **I**nformation **I**nterchange. It is a seven- or eight-bit code for representing alphanumerics, punctuation marks, and certain special characters for control purposes.

BAUD: The seven or eight bits that make up a character. A character can be a letter, number, symbol, etc.

BAUD RATE: A unit of data transmission speed equal to the number of code elements (characters) per seconds. For example, 300 baud is thirty characters—letters, numbers, symbols—per second. 1200 baud is 120 characters per second.

BINARY CODED DECIMAL (BCD): One of several numbering systems used with PLCs. This unique numbering system uses four binary digits to represent each decimal digit from 0 to 9. Groups of four binary digits are grouped together to display decimal numbers. Twelve bits can represent a three-digit number. Sixteen bits are needed to represent a four-digit number.

BINARY: A numbering system that uses a base of two. There are two digits (1 and 0) in the binary system.

BIT: An acronym for **B**inary dig**IT**. A bit can be only one of two possible states: *ON* or *OFF*; high or low; logic 1 or logic 0; etc.

BOOLEAN ALGEBRA: Shorthand notation for expressing logic functions.

BOOLEAN EQUATION: Expression of relations between logic functions and/or elements.

BRANCH: A parallel logic path within a user program rung.

BOOT: A term used in the computer world to indicate that a computer has been turned on, and the software has been loaded.

BREAKDOWN VOLTAGE: The voltage at which a disruptive discharge takes place, either through or over the surface of insulation.

BUFFER: A temporary storage area where information is held while the printer or other device catches up with the transmission speed of the data.

BYTE: A sequence of binary digits usually operated upon as a unit (normally eight bits).

CASCADING: A programming technique that extends the ranges of timer and/or counter instructions beyond the maximum values that normally may be accumulated.

CASSETTE RECORDER: A peripheral device for transferring information between PLC memory and magnetic tape.

CENTRAL PROCESSING UNIT (CPU): Another term for processor.

CHARACTER: One symbol of a set of elementary symbols, such as a letter of the alphabet, a decimal numeral, a punctuation mark, etc.

CHILD: A term used to identify a subdirectory in a main directory.

CLOCK: A device (usually a pulse generator) that generates periodic signals for synchronization or timing.

CMOS: An acronym for **C**omplimentary **M**etal **O**xide **S**emiconductor. A family of very low power, high-speed integrated circuits.

CODE: A system of symbols (bits) for representing data (characters).

COMPARE FUNCTION: A program instruction that compares numerical values for "equal," "less than," "greater than," etc.

COMPATIBILITY: The ability of various specified units to replace one another, with little or no reduction in capability.

COMPUTER GRADE TAPE: A high quality magnetic digital recording tape which must be rated at 1600 FCI (flux changes per inch), or greater.

COMPUTER INTERFACE: A device designed for data communication between a central computer and another unit, such as a PLC processor.

CONTACT SYMBOLOGY DIAGRAM: Commonly referred to as a ladder diagram, it expresses the user-programmed logic of the controller in relay-equivalent symbols.

COUNTER: A device that can count up or down in response to transitions (*OFF* or *ON*) of an input signal and opens and/or closes contacts when a predetermined count is reached. Counters are internal to the processor and are not real-world devices.

CPU: An abbreviation for **C**entral **P**rocessing **U**nit. It is used interchangeably with processor.

CRT: The **C**athode **R**ay **T**ube, which is an electronic display tube similar to the familiar TV picture tube. It is more commonly called a **VDT**, or **V**ideo **D**isplay **T**erminal.

CURSOR: A means for indicating on a VDT screen the point at which data entry or editing is to occur.

DATA MANIPULATION: The process of altering and/or exchanging data between storage words.

DATA TRANSFER: The process of exchanging data between PLC memory words and/or areas.

DECREMENT: A term used with counters to indicate that the value of the counter has decreased. When a counter value goes from 4 to 3, it is said to have decremented by 1.

DEFAULT: The initial setting of a value, or the initial assignment of a file by the software. Default values may or may not be changed, depending on the software and the PLC manufacturer.

DIGITAL: The representation of numerical quantities by means of discrete numbers. It is possible to express in binary digital form all information stored, transferred, or processed by dual-state conditions, e.g., *ON/OFF*, open/closed, etc.

DISCRETE INPUTS: An input that is either *ON* or *OFF*. Examples of discrete inputs are limit switches, push buttons, float switches, etc.

DISCRETE INPUT MODULE: A module that converts signals from real-world input devices to logic level signals for use by the processor.

DISCRETE OUTPUTS: An output that is either *ON* or *OFF*. Examples of discrete outputs are solenoids, motor starter coils, pilot lights, etc.

DISCRETE OUTPUT MODULE: A module that converts the logic levels of the processor to an output signal to control a real-world output.

DISKETTE: A magnetic medium for storing information that can later be read by the computer or PLC.

DOS®: **D**isk **O**perating **S**ystem. A registered trademark of the Microsoft Corporation. The full name is MS-DOS for **M**icro**S**oft **D**isk **O**perating **S**ystem.

DUMP: A term used when information stored in memory is copied or recorded onto magnetic tape or disk.

DUPLEX: A means of two-way data communication. Also see **FULL DUPLEX** and **HALF DUPLEX**.

EAROM: A type of programmable memory that can be erased or altered electrically. The term stands for **E**lectrically **A**lterable **R**ead **O**nly **M**emory.

EEPROM or **E²PROM:** A memory chip that can be programmed using a standard programming device, and can be erased when the proper signal is applied to the erase pin. The initials stand for **E**lectrically **E**rasable **P**rogrammable **R**ead **O**nly **M**emory.

ELECTRICAL-OPTICAL ISOLATOR: A device that couples different voltage levels using a light source and detector in the same package. It is used to provide electrical isolation between line voltage input and output circuitry and the processor.

ELEMENT: A program instruction (N.O. contact, timer, counter, etc.) displayed on a VDT.

ENABLED: A term used to indicate that a function or operation has been activated.

EVEN PARITY: The condition that occurs when the sum of a string of binary digits, 1s and 0s, add up to an even number. Parity is used for error checking.

EXAMINE OFF: An EXAMINE OFF PLC instruction is a true precondition if its addressed bit is *OFF* (0). It is false if the bit is *ON*, or 1.

EXAMINE ON: An EXAMINE ON instruction is a true precondition if its addressed bit is *ON* (1). It is false if the bit is *OFF*, or 0.

FALSE: When relating to PLC instructions, an *OFF* state or condition.

FAULT: Any malfunction which interferes with normal operation.

FILE: A group of words, usually consecutive, that are used to store information.

FORCE: A mode of operation or instruction that allows the operator (as opposed to the processor) to control the state of an input or output device.

FORCE OFF FUNCTION: A feature that allows the user to deenergize any input or output by means of the programmer, independent of the PLC program.

FORCE ON FUNCTION: A feature that allows the user to energize any input or output by means of the programmer, independent of the PLC program.

FORTRAN: An acronym for **FOR**mula **TRAN**slation, a scientific programming language.

FULL DUPLEX (FDX): A mode of communications in which data may be simultaneously transmitted and received by both ends (sender/receiver).

GROUND: A conducting connection, intentional or accidental, between an electric circuit or equipment chassis and the earth ground.

GROUND POTENTIAL: Zero voltage potential with respect to earth ground.

HALF DUPLEX (HDX): A mode of data transmission capable of communicating in two directions, but in only one direction at a time.

HARD CONTACTS: Any type of physical switch contacts. Contrasted with electronic switching devices, such as triacs and transistors.

HARD COPY: Any form of printed document such as a ladder diagram, program listing, data table configuration, etc.

HARDWARE: The mechanical, electrical, and electronic devices that comprise a programmable logic controller and its application.

HARDWARE KEY: A piece of hardware that is required for a program to run. It may be in the form of a plug or connector plugged into the printer port that allows the software to run. Without the hardware key installed, the software does not work, or limits the access to certain portions of the program.

HARD DRIVE: A storage system that consists of an inflexible (hard) disk, as opposed to a floppy disk, that is used to store files, directories, software programs, etc.

HARD-WIRED: Electrical devices interconnected through physical wiring.

HEXADECIMAL: The numbering system that represents all possible statuses of four bits with sixteen unique digits (0–9 then A–F).

HIGH = TRUE: A signal type wherein the higher of two voltages indicates a logic state of 1, or *ON*. (See **LOW = TRUE**.)

HOLDING REGISTER: A register or file that holds a value or values for comparison or for use in a user program.

IEEE: An acronym for **I**nstitute of **E**lectrical and **E**lectronics **E**ngineers.

IMAGE TABLE: An area in PLC memory dedicated to I/O data. Ones and zeros (1s and 0s) represent *ON* and *OFF* conditions, respectively. During every I/O scan, each input controls a bit in the *input image table*; and each output is controlled by a bit in the *output image table*.

INCREMENT: A term used with counters to indicate that the value has increased. When a counter has counted up from 3 to 4, it is said to have incremented by 1.

INPUT DEVICES: Devices such as limit switches, pressure switches, push buttons, etc., that supply data to a programmable logic controller. These real-world inputs are of two types: those with common returns, and those with individual returns (referred to as isolated inputs). Other inputs include analog devices and digital encoders.

INSTRUCTION: A command or order that causes a PLC to perform one certain prescribed operation.

INTERFACING: Interconnection of a PLC with its input and output devices and data terminals through various modules and cables. Interface modules convert PLC logic levels into external signal levels, and vice versa.

INTERPOSING RELAY: A relay that is added to a PLC circuit to handle current values larger than can be handled by one terminal of an output module.

INTERRUPTIBLE: Interruptible refers to a timer that can be interrupted, but still retain its accumulated time.

I/O: An abbreviation for Input/Output. For example, a group of input modules and output modules would be referred to as I/O modules.

I/O ELECTRICAL ISOLATION: Separation of the field-wiring circuits from the logic level circuits of the PLC. This is typically achieved using electrical-optical isolators mounted in the I/O module.

I/O MODULE: The printed circuit assembly that interfaces between the user devices and the PLC.

I/O RACK: A chassis which contains I/O modules.

I/O SCAN TIME: The time required for the PLC to monitor all inputs, read the user program, and control all outputs. The I/O scan repeats continuously.

I/O SECTION: Interfaces the different signals from real-world devices and sensors to signals the CPU can use.

JUMPER: A short length of conductor used to make a connection between terminals, around a break in a circuit, or around a device.

KEYED: A term used with PLCs to indicate that a keying device has been installed to prevent I/O modules from being installed in the wrong slot.

KILO: A prefix used with units of measurement to designate quantities 1000 times as great (as in kilowatt). The exception to K having a value of 1000 is when referring to computer memory. Computer memory is counted using the binary numbering system and one kilo (or K) of memory is 1024, not 1000. ($2^{10} = 1024$).

LADDER DIAGRAM: A complete control scheme normally drawn as a series of contacts and coils arranged between two vertical supply lines so that the horizontal lines of contacts appear similar to rungs of a ladder. A ladder diagram is normally the reference document used by the operator when entering the control program. (See **CONTACT SYMBOLOGY DIAGRAM**.)

LADDER DIAGRAM PROGRAMMING: A method of writing a user's PLC program in a format similar to a relay ladder diagram.

LANGUAGE: A set of symbols and rules for representing and communicating information (data) among people, or between people and machines.

LATCH: A device that continues to store the state of the input signal after the signal is removed. The input state is stored until the latch is reset.

LATCH INSTRUCTION: A PLC instruction which causes an output to stay *ON*, regardless of how briefly the instruction is enabled. (It can only be turned *OFF* by an UNLATCH INSTRUCTION in a separate Rung.)

LATCHING RELAY: A relay constructed so that it maintains a given position by mechanical means until released mechanically or electrically.

LEAST SIGNIFICANT DIGIT (LSD): The digit that represents the smallest value. In the number 102, the 2 is the least significant number, or digit.

LED: Acronym for **L**ight **E**miting **D**iode.

LIMIT SWITCH: A switch that is actuated by some part or motion of a machine or equipment to alter the electrical circuit associated with it.

LIQUID CRYSTAL DISPLAY (LCD): A reflective visual readout. Because its segments are displayed only by reflected light, it has extremely low power consumption, as contrasted with a LED display which emits light.

LOAD: 1) The power delivered to a machine or apparatus. 2) A device intentionally placed in a circuit or connected to a machine or apparatus to absorb power and convert it into the desired useful form. 3) To place data (e.g., a ladder diagram) into the processor's memory.

LOGIC LEVEL: The voltage magnitude associated with signal pulses representing ones and zeros (1s and 0s) in binary computation.

LOW = TRUE: A signal type wherein the lower of two voltages indicates a logic state of 1, or *ON*. (See **HIGH = TRUE**.)

MAGNETIC TAPE: Tape made of plastic and coated with magnetic material; used to store information.

MALFUNCTION: Any incorrect functioning within electronic, electrical, or mechanical hardware. (See **FAULT**.)

MANIPULATION: The process of controlling bits or words within a program to obtain the required program outcomes.

MASK: Bits in a word that are used to prevent other bits in a different word from being used. If there is a 1 in the bit location of the mask, the corresponding bit in the output word is enabled and can be turned *ON* and *OFF*. If the bit is set to 0, the corresponding bit in the output word is disabled, and does not allow the output to be turned *ON*, even if the program called for the bit to be turned *ON*.

MECHANICAL DRUM CONTROLLER: A type of SEQUENCER which operates switches by means of pins or cams placed on a rotating drum. The switch sequence may be altered by changing the pin or cam pattern.

MEMORY: The section of the programmable logic controller that stores the user program and other data. The storage may be either temporary or semipermanent.

MEMORY PROTECT: The hardware capability to prevent a portion of the memory from being altered by an external device. This hardware feature can be under actual key and lock control, or may use passwords that are referred to as software keys.

MENU: A display on the computer or PLC monitor that offers options or gives the operator choices to select from.

MIDDLE DIGIT (MD): The middle digit of a three digit number.

MILLIAMPERE (mA): One thousandth of an ampere: 10^{-3} or 0.001 ampere.

MILLISECOND (ms): One thousandth of a second; 10^{-3} or 0.001 second.

MINI-PLC: A scaled-down version of a standard PLC with small I/O capability.

MODE: A selected method of operation (for example, *run*, *test*, or *program*.)

MODEM: Acronym for **MO**dulator/**DEM**odulator. A device used to transmit and receive data by frequency-shift-keying (FSK). It converts FSK tones into their digital equivalents, and vice versa.

MODULE: An interchangeable "plug-in" item containing electronic components which may be combined with other interchangeable items to form a complete unit.

MOST SIGNIFICANT DIGIT (MSD): The digit representing the greatest value. In the number 102, the 1 is the most significant number, or digit.

MOTOR CONTROLLER: A device, or group of devices, that serves to govern, in a predetermined manner, the electrical power delivered to a motor.

NEMA STANDARDS: Consensus standards for electrical equipment approved by the majority of the members of the National Electrical Manufacturers Association.

NESTING: A programming technique that has a "branch-within-a-branch." Depending on the PLC manufacturer, nesting may or may not be allowed.

NETWORK: A group of connected logic elements used to perform a specific function. A network can range from one element to a complete matrix of elements, plus coil(s) as desired by the user. The size and configuration of the matrix or rungs varies with PLC manufacturers.

NODE: A common connection point between two or more contacts or elements in a circuit.

NOISE: Extraneous signals; any disturbance which causes interference with the desired signal or operation.

NOISE SPIKE: Voltage or current surge produced in the industrial operating environment.

NONVOLATILE MEMORY: A memory that is designed to retain its information even though its power supply is turned off.

NOT LOGIC: Not logic has the same logic as normally closed contacts. NOT logic means not open, or closed.

OCTAL NUMBERING SYSTEM: A numbering system that uses a base eight. Only the digits 0 through 7 are used.

ODD PARITY: The condition that occurs when the sum of 1s and 0s in a binary word is an odd number. Parity is used for error checking.

OFF DELAY TIMER: 1) In relay panel application, a device in which the timing period is initiated upon deenergization of its coil. 2) In a PLC, an instruction which starts the delay whenever the timer rung goes false.

OFF-LINE PROGRAMMING: A method of programming that is done while the processor is not communicating with the outputs.

ON DELAY TIMER: 1) In relay panel applications, a device in which the timing period is initiated upon energization of its coil. 2) In a PLC, an instruction which starts the delay whenever the timer rung goes true.

ON-LINE OPERATION: Operations where the programmable logic controller is directly controlling the machine or process.

ON-LINE PROGRAMMING: A method of programming by which rungs in the program may be inserted, changed, or deleted while the processor is running and controlling the process equipment.

OPERAND: 1) Either of the two numbers used in a basic computation to produce an answer. For example, in the computation 2 X 3 = 6, 2 and 3 are the operands. 2) Data required for the operation of a special function.

OPTICALLY COUPLED/OPTICAL ISOLATION: The use of a light emitting diode and a photo transistor to communicate a signal or state to the processor. Optical coupling is used in input and output modules to isolate the logic level signal for line voltage sources.

OR LOGIC: Logic that has a "parallel" relationship. When two devices are in parallel, if either device one OR device two is true, the logic passes.

OUTPUT CIRCUIT: An output module point, real-world device (e.g., motor starter, digital readout, solenoid, etc.), and its associated wiring. The output module's function is to convert processor signal levels to field voltage levels necessary to control the real-world devices.

OUTPUT DEVICES: Devices such as solenoids, motor starters, etc., that receive data (control) from the programmable logic controller.

OUTPUT SIGNAL: A signal provided by the processor to the real-world output devices that control their status (*ON* or *OFF*).

OVERLOAD: A load greater than that which a device is designed to handle.

PARALLEL COMMUNICATIONS: A type of communication or information transfer whereby a group of digits (bytes) are transmitted simultaneously. This is different from serial communications where the data bits are transmitted one at a time—sequentially—in a string.

PARENT: A name given to a directory that has subdirectories.

PARITY: A method of testing the accuracy of binary numbers used in recorded, transmitted, or received data.

PARITY BIT: An additional bit added to a binary word to make the sum of the number of 1s in a word always even or odd.

PARITY CHECK: A check that tests whether the number of 1s in an array of binary digits is odd or even.

PASSWORD: A word used to gain access to a program or process. A password serves the same function as a hardware key, except that it is not a piece of hardware, but rather a word that when entered at the keyboard, gains access for the user to the program.

PC: Abbreviation for **P**ersonal **C**omputer (also used as an abbreviation for **P**rogrammable **C**ontroller). To avoid the confusion caused by using the same abbreviations for two different types of systems, the programmable logic controller is now most often referred to as a PLC.

PLC: See PROGRAMMABLE LOGIC CONTROLLER.

PILOT DEVICE: A device used in a circuit that performs a control function only. Pilot devices are limited to 10 amps of current carrying capacity, and are to be used in control circuits only. They are not designed to control the power and current required by the operating equipment.

POWER SUPPLY: Supplies the DC power for the CPU and for the I/O section. The voltage is typically +5 volts. The power supply can be internal with the processor, rack mounted, or externally mounted as a separate unit. A separate power supply is required if DC voltage is required for the actual input and/or output devices.

PRIORITY: Order of importance.

PROCESSOR/PROCESSOR UNIT: The part of the programmable logic controller that performs logic solving, program storage, and special functions within a PLC system. It scans all the inputs and outputs in a predetermined order. The processor monitors the status of the inputs and outputs in response to the user programmed instructions, and energizes or deenergizes outputs as a result of the logical comparisons made through these instructions.

PROGRAM: A sequence of instructions executed by the processor to control a machine or process.

PROGRAM PANEL (PROGRAMMER): A device for inserting, monitoring, and editing a program in a programmable logic controller.

PROGRAM SCAN TIME: The time required for the processor to execute all instructions in the program one time.

PROGRAMMABLE LOGIC CONTROLLER (PLC): A solid-state control system that has a user programmable memory for storage of instructions to implement specific functions such as I/O control logic, timing, counting, arithmetic, and data manipulation. A PLC consists of the processor, input/output interface, memory, and programming device that typically uses relay-equivalent symbols. The PLC is purposely designed as an industrial control system to perform functions equivalent to a relay panel or a wired solid-state logic control system.

PROGRAMMER: A device that is needed to enter, modify, and troubleshoot the PLC program, and check the condition of the processor. The programmer may be hand-held, dedicated desktop-type, or a personal computer.

PROM: Acronym for **P**rogrammable **R**ead **O**nly **M**emory. A type of read only memory that requires an electrical operation to generate the desired bit or word pattern. In use, bits or words can be accessed on demand, but cannot be changed.

PROTECTED MEMORY: Storage (memory) locations reserved for special purposes or use by the processor into which data cannot be entered directly by the user.

PROTOCOL: A defined means of establishing criteria for receiving and transmitting data through communication channels.

RS-232C: An Electronic Industries Association (EIA) standard for data transfer and communication.

RACK: A PLC chassis that contains modules. Some PLC manufacturers, like Allen-Bradley, use the term rack to indicate a given number of I/O points rather than to identify a specific piece of hardware. In the Allen-Bradley scheme, a chassis could contain a number of racks. With most other manufacturers, however, rack and chassis are used interchangeably, and mean the hardware that holds the various modules, power supplies, etc.

RAM: Acronym for **R**andom **A**ccess **M**emory. Random access memory is a type of memory that can be read from (accessed) or written into by the user.

RANDOM ACCESS: See **RAM.**

RATED VOLTAGE: The maximum voltage at which an electrical component can operate for extended periods without damage or undue degradation.

READ/WRITE MEMORY: A memory into which data can be placed (*write* mode) or accessed (*read* mode). The *write* mode destroys previous data; the *read* mode does not alter stored data.

REGISTER: A word or group of words used to store numerical values.

REPORT: A display of data, or a printout, containing data and/or information that is useful to the user or operator. Reports can include operator messages, part records, production lists, etc. Reports are normally stored in a memory area separate from the user's program.

REPORT GENERATION: The printing or displaying of user-formatted application data by means of a programming device. Report generation can be initiated by means of either the user's program or a programming device keyboard.

RETENTIVE: To retain a value or time.

RETENTIVE OUTPUT: An output that remains in its last state (*ON* or *OFF*), depending on which of its two program Rungs (one containing a LATCH INSTRUCTION, the other an UNLATCH INSTRUCTION) was the last to be true. The retentive output remains in its last state when both Rungs are false. It also remains in its last state if power is removed from, then restored to, the PLC.

RETENTIVE TIMER: A PLC instruction which accumulates the amount of time, whether continuous or not, when the preconditions of its rung are true, and which controls one or more outputs after the total accumulated time is equal to the preset time. When the rung is false, the accumulated time is retained. Moreover, if the outputs have been energized, they remain *ON*. Additionally, the accumulated time and energized outputs are retained if power is removed from, then restored to, the PLC.

ROM: Acronym for **R**ead **O**nly **M**emory. A read only memory is a solid-state digital storage memory whose contents cannot be altered by the user.

RUNG: A grouping of PLC instructions that control one output or storage bit. Some PLCs can have multiple outputs on the same rung. A rung is also referred to as a network.

SCAN: The time required to make one complete scan through memory and update the status of all inputs and outputs.

SCHEMATIC: A diagram of a circuit in which symbols illustrate circuit components.

SCR: An acronym for **S**ilicon **C**ontrolled **R**ectifier. The SCR is used to convert AC current to DC current.

SEQUENCER: A controller which operates an application through a fixed sequence of events.

SERIAL COMMUNICATION: A type of communication or information transfer within a programmable logic controller whereby the bits are handled sequentially rather than simultaneously as they are in parallel communications. Serial operation is slower than parallel operation for equivalent clock rate. However, only one channel is required for serial operation.

SHIELDING: The practice of confining the electrical field around a conductor to the primary insulation of the cable by putting a conducting layer around the cable insulation.

SOFTWARE: The manufacturers' program that controls the operation of a programmable logic controller.

SOLID-STATE: Circuitry designed using only integrated circuits, transistor, diodes, etc.; no electromechanical devices such as relays are utilized. High reliability is obtained with solid-state logic.

SOLID-STATE DEVICES (SEMICONDUCTORS): Electronic components that control electron flow through solid materials (e.g., transistors, diodes, integrated circuits, etc.).

STATE: The logic condition, 1 or 0, in PLC memory, or at a circuit's input or output.

STORAGE: Synonymous with MEMORY.

STORAGE MEMORY: That part of the memory that stores the status of the input and output devices, numeric values for timers and counters, numeric values for arithmetic functions, status of internal relays, and information stored in holding and storage registers.

SURGE: A transient variation in the current and/or voltage at a point in the circuit.

SWITCHING: The action of turning a device *ON* and *OFF*.

SYMBOLIC NAME: A user designation for an application I/O device (e.g., S-1, LS-4, or SOL-7).

SYSTEM PROMPT: The system prompt indicates the current drive for the computer. If the prompt is C:\, then the current drive is C drive.

THUMBWHEEL SWITCH: A rotating numeric switch used to input numeric information to a controller.

TIMER: In relay-panel hardware, an electromechanical device that can be wired and preset to control the operating interval of other devices. In a PLC, a timer is internal to the PROCESSOR, meaning that it does not exist in the real world, but can be controlled by a user-programmed instruction. A timer instruction has greater accuracy and timing range than a hardware timer.

TOGGLE SWITCH: A panel-mounted switch normally used for *ON* or *OFF* switching.

TRANSFORMER COUPLING: One method of isolating I/O devices from the controller.

TREE: A command that is used to display the organization of all the directories, subdirectories, and files on a given disk or hard drive.

TRIAC: A solid-state component capable of switching alternating current.

TRUE: As related to a PLC instruction, an enabling logic state or *ON* condition. (See FALSE.)

TRUTH TABLE: A matrix which shows all the possible states (*ON* or *OFF*) of a single input device or combination of input devices, and the corresponding state (*ON* or *OFF*) of the output device.

TTL: Abbreviation for **T**ransistor-**T**ransistor **L**ogic, a family of integrated circuit logic. (Usually 5 volts is high, or 1, and 0 volts is low, or 0.)

UPDATING: A term used to indicate that the processor has scanned and checked the status of all input and output devices. After the status of the inputs and outputs are known, the data table is updated to reflect the current status.

UPWARD COMPATIBILITY: The ability of a new version of software to support previous editions of the same software. If version 6.6 of a particular software allows documents and files from previous versions (5.2 and 6.0) to be read, the software is said to have upward compatibility.

UNLATCH INSTRUCTION: A PLC instruction which causes an output to unlatch, or turn *OFF*, regardless of how briefly the instruction is enabled. (It can only be turned back *ON* by a LATCH INSTRUCTION in a separate Rung.)

UNCONDITIONAL: A term applied to an output (or other instruction) that is always true.

USER MEMORY: The portion of memory that is set aside for the storage of the user program (i.e., ladder diagrams, program messages, etc.).

UV ERASABLE PROM: An erasable programmable read only memory which can be erased or cleared (set to 0) by exposure to intense ultraviolet light. After being cleared, it may be reprogrammed.

VALUE: 1) A number that represents a computed or assigned quantity. 2) A number contained in a register or file word.

VOLATILE MEMORY: A memory that loses its information if the power is removed.

WATCHDOG TIMER: A timer that is used within the PLC processor to verify that the program scan has been completed correctly in the allotted amount of time. If there is a program error, the scan is not completed in the prescribed amount of time, and the watchdog timer times out and indicates that there is a problem with the circuit.

WORD: A grouping, or a number of bits, in a sequence that is treated as a unit.

WRITING OVER: A term used to indicate that information will be replaced (or the existing information will be written over) by new information.

INDEX

A

Accumulated time, 196-97
ADD instruction, 268
Addition
  (ADD) instruction, 268
  binary numbers, 278
  binary table, 276-77
  decimal system of, 275, 276
Addressing
  scheme, PLC-5, 146-60
  words, 74
Algebra, Boolean, 310-14
Allen-Bradley
  counters, 225-28
    PLC-5, 228-34
  PLC-2 timers, 196-204
  PLC-5
    addressing scheme, 146-60
    data compare instructions, 257-63
    data transfer instructions, 248-50
    file structure, 84-86
    math functions, 269-70
    timers, 204-11
Always false instruction, 187-88
Analog I/O modules, 35-43
  grounding, 40
  rack installation, 35, 37-39
  safety circuit, 35
  shielding, 41-43
  surge suppression, 39-40
Analog input devices, 7
AND logic, 311
  programming, 316, 318-20
Arithmetic math functions
  Allen-Bradley PLC-2, 268-69
  Gould, 271-75
ASCII, 98-100
Asynchronous shift register
  FIFO, 298-302
  LIFO, 302
    stack command, 318

B

BCD (Binary coded decimal system), 97-98
Binary
  addition table, 276-77
  coded decimal (BCD) system, 97-98
  digit, defined, 73

numbers
  addition of, 278
  converting to decimal equivalents, 278
system, 90-92
  described, 54
Bit shift, 285-90
Bits
  described, 73
  Done (DN), 205
  enable (EN), 205
BLKM instruction, 296-97
Block move, 296
  instruction, 296-97
Blown-fuse indicators, 27
Boole, George, 310
Boolean algebra, 310-14
  programming in, 314-23
  see also Truth table
Booted, defined, 328
Branch circuit
  end, 126
  start, 126
Break command, 335
Buffer, described, 161

C

Cascading timers, 221-22
Cassette recorders, 59
CD (Change directory) command, 337
CD (Count down enable bit), 229
Change directory command, 337
Chassis, 14-15
Child directory, described, 332
Circuits
  one shot, 245
  safety, 181-82
  STOP/START, 106-8
Clear screen command, 335
CMOS-RAM, 52
CMP instruction, 259-61
Cold started, defined, 328
Colon, use of, 147
Combination circuit, programming, 322
Commands
  break, 335
  change directory, 337
  clear screen, 335
  copy, 335-36

Commands, *(continued)*
  delete files, 338
  DIR, 333-34
  directory tree, 338-40
  FORMAT, 329-30
  interrupt, 335
  make directory, 337
  PRINT, 340
  remove directory, 338
  renaming files, 337
  STR, 318
Communication
  parallel, 160
  serial, 160
Compare instruction, 259-61
Compliment, 2s, 275-83
Computer
  (CPT) instruction, 269-70
  programmers, 66-70
  programming with, 138-46
  starting, 328-29
  troubleshooting, 346-50
Contact output modules, 33-34
Contacts
  discrete holding, 172
  nesting, 168
  overload, 172-74
Continuous duty rating, 24
Control in, 218
COP instruction, 297-98
Copy command, 335-36
Count down
  done bit (DN), 229
  enable bit (CD), 229
  underflow bit (UN), 229
Count up
  done bit (DN), 229
  enable bit (CU), 229
  overflow bit (OV), 229
Counters
  Allen-Bradley, 225-28
    PLC-5, 228-34
  Modicon, 234-37
  Square D Company, 237-39
  timers combining with, 239
CPT instruction, 269-70
CU (Count up enable bit), 229
Cursor, function of, 130

D
Data
  comparisons
    PLC-5 instructions, 257-63
    programmed in parallel, 255
    Square D Company, 252, 253
    time chart for, 254
  table, 80
    words, 242
  transfer, 242-48
    instructions, 250
      PLC-5, 248-50
Decimal
  addition system, 275, 276
  numbers, converting binary numbers to, 278
  system, 88-89
Dedicated desktop programmers, 62-65
  advantages/disadvantages of, 71
Dedicated tape loader, 59
Default drive, described, 328
DEL command, 338
Delete files command, 338
Desktop programmers
  advantages/disadvantages of, 71
  dedicated, 62-65
Diagrams
  ladder, 103
    rules, 104-6
    sequenced motor starting, 108-9
    STOP/START circuit, 106-8
  wiring, 102
DIP switch, 15
Direct current, output modules, 25-30, 30-33
Directories, function of, 331
Directory
  creation, MS-DOS and, 330-33
  tree command, 338-40
Discrete
  holding contacts, 172
  I/O modules, 16-34
Discrete devices
  input, 7
  input modules, 16-17
    alternating current, 17-22
    direct current, 22-23
  output, 7
  output modules, 23-25
    alternating current, 25-30

direct current, 30-33
Disk formatting, 329-30
Display terminals, video, 64-65
DIV instruction, 269
Divide (DIV) instruction, 269
DN
    (Count down done bit), 229
    (Count up done bit), 229
    (Done bits), 205, 229
Done bit, 205
    count down, 229
    count up, 229
Double precision arithmetic, 268
Drum cylinder, 304
Dual-in-line package, 15
Dummy relay, 163-64
DX instruction, 250

E
EAROM, 53
EEPROM, 53-54
Electrical noise, analog I/O modules and, 39-40
Electrically Alterable Read Only Memory, 53
Electrically Erasable Programmable Read Only
    Memory, 53-54
Electromagnetic interference (EMI), described,
    39
EMI (Electromagnetic interference), 39
EN (Enable bits), 205, 229
Enable bit, 205
    count down, 229
    count up, 229
Enable/set, 218
EQU instruction, 258
Examine off, 116-21
    clarifying, 121-24
Examine on, 116
    clarifying, 121-24

F
Fast-responding DC input modules, 23
FFL instruction, 299-300
FFU instruction, 299-300
FIFO asynchronous shift register, 298-302
File
    copy instructions, 297-98
    creation, MS-DOS, 330-33
    moves, 290

names, 141
    MS-DOS and, 330
    structure, PLC-5, 84-86
File-to-file instruction, 292-98
File-to-word instruction, 291
Final system checkout, 345-46
First out (FOUT) function, 301
First-in-first out (FFL) instruction, 299-300
First-in-first out unload (FFU) instruction,
    299-300
Fixed I/O, 12-13
FORMAT command, 329-30
Formatting, disk, 329-30
FOUT function, 301
Freeze pointer line, 293
Functions, FOUT, 301
Fuses, output, 27-28

G
Gould, math instruction, 271-75
Grounding, analog I/O modules, 40

H
Hand-held programmers, 65-66
    advantages/disadvantages of, 71
Heat sinks, 24
Hexadecimal system, 94-97
Holding contacts, 107
    discrete, 172
Holding register, 242

I
I/O
    fixed, 12.3
    modules, 35-43
        discrete, 16-34
        grounding, 40
        rack installation, 35, 37-39
        safety circuit, 35
        shielding, 41-43
        surge suppression, 39-40
        shielding, 41-43
IF instruction, 267
Immediate input instruction, 182-83
Immediate output instruction, 183-84
Input
    devices
        analog, 7

Input, *(continued)*
    discrete, 7
    instructions, immediate, 182-83
Input modules
    analog, 35-43
        grounding, 40
        rack installation, 35, 37-39
        safety circuit, 35
        shielding, 41-43
        surge suppression, 39-40
    discrete, 16-17
        alternating current, 17-22
        direct current, 22-23
Input/output section, 6-7
    described, 11-12
    fixed, 12-13
    modular, 14-16
    modules
        discrete, 16-34
        input, 16-17
Inputs, testing, 343
Instruction set, 146
Instructions
    ADD, 268
    always false, 187
    BLKM, 296-97
    block move, 296-97
    CMP, 259-61
    COP, 297-98
    CPT, 269-70
    data transfer, 250
    DIV, 269
    EQU, 258
    FFL, 299-300
    FFU, 299-300
    file copy, 297-98
    file-to-file, 292-98
    file-to-word, 291
    GRT, 258
    IF, 267
    immediate
        input, 182-83
        output, 183-84
    jump, 185
        to subroutine, 186
    label, 185
    latching relay, 177, 178-81
    LEQ, 259

LET, 266
LIM, 262-63
logical holding, 172
master control relay, 176-78
MOV, 248-49
MUL, 269
MVM, 249-50
one shot, 187-88
relay type
    examine off, 116-24
    examine on, 116, 121-24
    programming contacts, 113-16
return, 186
sequencer, 304
SUB, 268
subroutine, 186
temporary end, 186
word-to-file, 290-91
Interface, defined, 3
Internal relays, 163-64
Interposing relay, 33-34
Interrupt command, 335
Isolation, electrical noise, 39

J
Jump instructions, 185
    to subroutine, 186

K
Keyboard, 62-64
Keying, module, 29-30

L
Label instructions, 185
Ladder diagrams, 103
    rules, 104-6
    sequenced motor starting, 108-9
    STOP/START circuit, 106-8
Ladder programming logic, 315
Latching relay, instructions, 177, 178-81
LEQ instruction, 259
LET instruction, 266
LIFO
    asynchronous shift register, 302
    stack command, 318
Lights
    logic, 29
    output, 29

status, 28-29
LIM instruction, 262-63
Logic light, 29
Logical holding instructions, 172

M
Magnetic tape loader, 59
Maintaining contacts, 107
Make directory command (MD), 337
Mask word, 307
Masked move (MVM) instruction, 249-50
Masks, 307-9
Master control relay instructions, 176-78
Math functions
    Allen-Bradley
        PLC-2, 268-69
        PLC-5, 269-70
    Gould, 271-75
    using, 265-68
MD command, 337
Mechanical latching relay, 179-80
Memory
    nonvolatile, 50
    organization, 79-84
        words/word locations, 73-79
    size, 54-56
    storage, 50, 57, 80-82
    structure, 56-57
    types of, 52-54
    user, 50, 56, 82, 84
    volatile, 50
Menu, function of, 140
Microprocessor, described, 48
MODEM, 59
Modicon Inc.
    counters, 234-37
    timers, 218-21
Modular I/O, 14-16
Module keying, 29-30
Modules
    discrete input, 16-17
        alternating current, 17-22
        transistor-transistor logic (TTL), 34
Motor starting, sequenced, 103-9
MOV instruction, 248-49
Move (MOV) instruction, 248-49
MS-DOS
    commands

        change directory, 337
        clear screen, 335
        copy, 335-36
        delete files, 338
        DIR, 333-34
        directory tree, 338-40
        FORMAT, 329-30
        interrupt/break, 335
        make directory, 337
        PRINT, 340
        remove directory, 338
        renaming files, 337
    file/directory creation with, 330-33
    starting a computer and, 328-29
MUL instruction, 269
Multiply (MUL) instruction, 269
MVM instruction, 249-50

N
National Electrical Manufacturing Association
    (NEMA), definition of
        programmable controller, 2
Negative transition contact, 294
Nesting, contacts, 168
Network
    defined, 162
    limitation of, 162-65
    programming restrictions, 166-69
Noise, electrical, analog I/O modules and, 39-40
Nonvolatile memory, 50
NOT logic, 312
    programming, 316
Number systems
    binary coded decimal (BCD) system, 97-98
    binary system, 90-92
    decimal system, 88-89
    hexadecimal system, 94-97
    octal system, 92-94
    using, 98-100

O
Octal system, 92-94
Off delay timer, 195
Off line
    described, 68
    programming, 140
On delay, described, 191
On line programming, function of, 127, 140

One shot
    circuit, 245
    instructions, 187-88
Optical isolation, described, 17
OR logic, 310-11
    programming, 317, 318-20
Output
    devices, discrete, 7
    fuses, 27-28
    instructions, immediate, 183-84
    light, 29
    testing, 343-45
Output modules, 35-43
    contact, 33-34
    discrete, 23-25
        alternating current, 25-30
        direct current, 30-33
    grounding, 40
    rack installation, 35, 37-39
    reed relay, 34
    safety circuit, 35
    shielding, 41-43
    surge suppression, 39-40
OV (Count up overflow bit), 229
Overflow bit, count up, 229
Overload contacts, 172-74

P
Parallel
    communication, 160
    outputs, programming, 321
Parent directory, described, 332
Passwords, 127
Path, described, 332
Peripherals, 160-61
    described, 57-59
Personal computers, advantages/disadvantages
    of, 69-72
PLC-2 timers, 196-204
PLC-5
    addressing scheme, 146-60
    data compare instructions, 257-63
    data transfer instructions, 248-50
    file structure, 84-86
    math functions, 269-70
    timers, 204-11
Pneumatic timers, 190-96
Positive transition contact, 294

POST (Power-on self-test), 328
Power supply, 4-6
Power-on self-test (POST), 328
PR (Preset time), described, 196-97
Preset time (PR), described, 196
PRINT command, 340
Printers, 57
Processor unit, 3
    described, 47-51
Program scanning, 169-71
Programmable logic controller (PLC)
    components of, 3
    defined, 1, 2
    described, 1-8
    input/output section, 6-7
    power supply, 4-6
    processor unit, 3
        described, 47-51
    programming device, 7-8
    programming small, 324-25
Programmable read only memory (PROM),
    52
Programmed latch, 180
Programming
    AND logic, 316, 318-20
    combination circuit, 322
    with a computer, 138-46
    contacts, relay type instructions and, 113-16
    ladder logic, 315
    on line, 127, 140
    NOT logic, 316
    off line, 140
    OR logic, 317, 318-20
    parallel outputs, 321
    restrictions, network, 166-69
    shorthand method, 154
    small PLC, 324-25
    STOP buttons, 171-72
    timers, 323-24
Programming devices, 7-8
    computer, 66-70
    computer programmers, 69-72
    dedicated desktop, 62-65, 71
        advantages/disadvantages of, 71
    described, 61-62
    hand-held, 65-66
        advantages/disadvantages of, 71
        programmers, 65-66, 71

personal computers, advantages/disadvantages of, 69-72
Programs, sample, 127-38
PROM (Programmable read only memory), 52

R

Racks, 14-15
  installation, analog I/O modules, 35, 37-39
RAM (Random Access Memory), 52
Random Access Memory (RAM), 52
RD command, 338
Read only memory (ROM), 52
Real world, defined, 7
Reed relay output module, 34
Register
  holding, 242
  synchronous shift, 285-90
  table, 80
Relay
  dummy, 163-64
  internal, 163-64
  interposing, 33-34
  latching, instructions, 177, 178-81
  master control, instructions, 176-78
Relay type instructions
  examine off, 116-21
    clarifying, 121-24
  examine on, 116
    clarifying, 121-24
  programming contacts, 113-16
Remove directory command, 338
REN command, 337
Renaming files command, 337
Reset pointer line, 293
Retentive time, 203
  reset (RTR), 204
Return instruction, 186
ROM (Read only memory), 52
RTR (Retentive time reset), 204

S

Safety circuit, 181-82
  analog I/O modules, 35
Scan, described, 48
Scanning, program, 169-71
Sealing contacts, 107
Semicolon, use of, 147
Sequenced motor starting, 108-9

Sequencers, 304-9
  instruction for, 304
  table, 305
Serial
  communication, 160
  shift register, 285-90
Shielding, I/O, 41-43
Shift register, synchronous, 285-90
Shorthand method programming, 154
Single precision arithmetic, 268
Soft keys, described, 127-29, 132
Sourcing and sinking, described, 31-33
Square D Company
  compare format, 252
  counters, 237-39
  data comparisons, 253
  timers, 212-18
Start-up procedures, 341-43
Status lights, 28-29
STOP buttons, programming, 171-72
STOP/START circuit, 106-8
Storage
  memory, 50, 57, 80-82
  register
    data comparisons, 254
    words, 242
STR command, 318
SUB instruction, 268
Subdirectories, use of, 331-33
Subroutine instruction, 186
Suppression, electrical noise, 40
Surge
  rating, 24
  suppression, analog I/O modules and, 39-40
Synchronous shift register, 285-90
System
  checkout, final, 345-46
  prompt, described, 328

T

Tape loaders, 59
Temporary
  end instruction, 186
  file, entering data in, 317-18
Terminal, video display, 64-65
Testing
  inputs, 343
  outputs, 343-45

Timers
  Allen-Bradley
    PLC-2, 196-204
    PLC-5, 204-11
  cascading, 221-22
  counters combining with, 239
  Modicon Inc., 218-21
  pneumatic, 190-96
  programming, 323-24
  Square D Company, 212-18
  watchdog, 49
Total current rating, 24
Transfer, data. *See* Data transfer
Transistor-transistor logic (TTL) I/O module, 34
Transition contacts, 294
TREE command, 338-40
Triac, described, 25-26
Troubleshooting, 346-50
Truth table
  AND, 311
  AND, OR, and NOT, 314
  AND and NOT, 313
  NOT, 312
  NOT and OR, 313
  OR, 311
  STOP/START logic, 312

  see also Boolean algebra
TTL I/O module, 34
Twos compliment, 275-83

U
Ultraviolet Programmable Read Only
  Memory, 53
UN (Count down underflow bit), 229
Underflow bit, count down, 229
User memory, 50, 56, 82, 84
UVPROM-EPROM, 53

V
Video display terminals, 64-65
Volatile memory, 50

W
Watchdog timer, described, 49
Wiring diagrams, 102
Words
  addressing, 74
  data table, 242
  described, 285
  locations in memory, 73-79
  storage register, 242
Word-to-file instruction, 290-91